Bacteria: A Very Short Introduction

VERY SHORT INTRODUCTIONS are for anyone wanting a stimulating and accessible way in to a new subject. They are written by experts, and have been published in more than 25 languages worldwide.

The series began in 1995, and now represents a wide variety of topics in history, philosophy, religion, science, and the humanities. The VSI library now contains more than 300 volumes—a Very Short Introduction to everything from ancient Egypt and Indian philosophy to conceptual art and cosmology—and will continue to grow in a variety of disciplines.

Very Short Introductions available now:

For more information visit our website
www.oup.co.uk/general/vsi/

Sebastian G. B. Amyes

BACTERIA

A Very Short Introduction

OXFORD
UNIVERSITY PRESS

OXFORD
UNIVERSITY PRESS

Great Clarendon Street, Oxford, OX2 6DP,
United Kingdom

Oxford University Press is a department of the University of Oxford.
It furthers the University's objective of excellence in research, scholarship,
and education by publishing worldwide. Oxford is a registered trade mark of
Oxford University Press in the UK and in certain other countries

© Sebastian G. B. Amyes 2013

The moral rights of the author have been asserted

First Edition published in 2013

Impression: 1

British Library Cataloguing in Publication Data

Data available

ISBN 978-0-19-957876-4

Printed in Great Britain by
Ashford Colour Press Ltd, Gosport, Hampshire

For Hilary and Lucy

Acknowledgements

I should like to thank Dr H.-K. Young and Dr A. K. B. Rolfe for their very useful scientific and medical advice and for their valuable comments on the manuscript. I am also indebted to Emma Marchant at Oxford University Press for her help with the production of this book.

Contents

CONTENTS

Preface

It would be understandable if we thought that humans were the principal species on this planet and that we now live in the era where mammals dominate. As we consider previous ages ending with the Cretaceous extinction 65.5 million years ago, we could well be forgiven for thinking that this was the 'Age of the Dinosaurs'. The reason is that we tend to classify each era with what can easily be seen around us or from what palaeontologists have reported and placed in museums of natural history for us to marvel at. The truth is there never have been any dominant organisms other than bacteria and that this planet has been in the 'Age of Bacteria' almost from the very beginning when life emerged. Bacteria are the most numerous of all organisms and their biomass is by far the largest on our planet and has been estimated to be greater than all the rest combined. Even within our own bodies the number of bacterial cells outnumbers our own cells. They can survive almost everywhere on the planet, from the coldest to the hottest places on earth, even to the bottom of the oceans. No other organisms are as adaptable.

List of illustrations

List of illustrations

Chapter 1
Origins

It would be easy for us to assume that bacteria are the simplest form of life and thus presumably would have been the original life form on this planet. This may be true but it is not a simple equation. Bacteria are single-cell organisms and are what are known as prokaryotic cells. These differ considerably from the cells of both animals and plants inasmuch as there are no visible discrete compartments within the cell. They are also usually considerably smaller than the cells of animals and plants.

So how did bacteria first emerge? This looks like a classic 'chicken and egg' conundrum. Bacteria, like all cells, contain DNA and they function by the decoding of this DNA into proteins, which comprise the enzymes that control all the major processes within the organism. In this respect, they are similar to other cells and thus probably have a common origin. The link between DNA decoding and protein production is RNA. RNA does not differ greatly in structure from DNA and some believe that RNA is the origin of life. This is plausible as RNA is the messenger; it is the molecule that is transcribed from DNA and from which protein is translated. It was the discovery of ribozymes by Thomas Cech, at the University of Colorado, and Sydney Altman of Yale University that strongly suggested that RNA was the origin of life. Ribozymes are RNA molecules that have a 3D (tertiary) structure and they can act as catalysts, similar to enzymes. Therefore RNA could act not only as the store of genetic material but also as the 'enzyme' that decodes it into the structures of life.

We may never be able to confirm this hypothesis but if we assume that it is plausible, then we can start to examine how bacteria emerged and where they fit in the evolutionary tree. Approximately 4.3 billion years ago, the first cells are thought to have arisen, probably with RNA as an essential catalytic role and later as a self-replicating molecule. The basic integrity of a cell is the formation of a cell membrane, composed of lipid bilayers. As these can form spontaneously, they could have surrounded early RNA molecules. Their continued presence may have been promoted through mutation of the RNA, which would have been passed on to succeeding generations through self-replication. This basic system does have significant disadvantages because a mistake made in the replication of RNA would immediately have an effect not only on the replication of the genetic material but also on the ability to act as a catalyst—largely, it may be assumed, in a detrimental manner. The separation of the self-replicating machinery from the enzymes they encode would have resulted in far fewer abortive stages. Consequently, we must assume that DNA largely took over the role of the carrier for the self-replicating genes and proteins of the enzymes that they eventually encoded. RNA merely remained as the messenger that carried the instructions from DNA to the formation of the proteins.

The early bacteria emerged approximately 1.5 billion years after the creation of the planet. For the next three billion years, bacteria were probably the planet's sole living inhabitants. The fossil record shows that there were huge numbers of bacteria often collecting in large colonies, attached to many surfaces; their imprint can still be seen. Approximately one billion years ago, however, the numbers of these colonies in the fossil record began to fall and this could be evidence that bacteria had become the source of food for some other life form. Certainly, this was before the Cambrian explosion of half a billion years ago, when the diversity of life forms increased rapidly, but it does suggest that, for the first three-quarters of life on Earth, bacteria had it all their own way.

What were these bacteria?

The common view is that prokaryotic cells, such as bacteria, and eukaryotic cells, such as those that comprise our bodies, had a common ancestor. The last universal common ancestor (LUCA) or cenancestor is considered to be the most recent ancestor of *all* life on Earth and possibly was living some 3.5 billion years ago (Figure 1). It is thought to have been a prokaryotic, single-celled organism possibly similar to simple bacteria found today. By this time, it has been concluded that the genetic code must have become DNA rather than RNA, and that the catalysts had become true enzymes (proteins) composed of the twenty amino acids. Furthermore, the

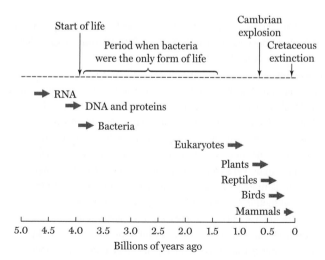

1. **Timescale of bacteria emergence**

machinery for dividing the DNA, maintaining its integrity, and expressing the genes through RNA was already established.

These bacteria would have been exclusively anaerobic; they did not respire oxygen as there was little or no available oxygen in the atmosphere at the time. These bacteria could produce energy from the available nutrients but this was an extremely inefficient process. About 3.2 billion years ago, photosynthetic bacteria or cyanobacteria first emerged. These bacteria could use energy from the sun to make sugars, which were used for further metabolism. The by-product of this photosynthesis was oxygen, which began to accumulate in the atmosphere. Oxygen is toxic to many cells, particularly the early anaerobic bacteria, which probably started to decline. About 2.5 billion years ago, the fossil record shows that aerobic bacteria emerged, able to use the newly available oxygen and use it to convert sugars into energy, usually in the form of adenosine-5'-triphosphate (ATP). The use of oxygen vastly increased the energy obtained from a single sugar molecule, and these bacteria soon predominated.

The main eukaryotic cells may have derived from an early example of Archaea bacteria, which themselves derived from LUCA. They probably evolved about 1.5 billion years ago. These were distinguishable from prokaryotic cells by having a defined nucleus, usually comprising discrete chromosomes, that contained the DNA. However, having emerged from bacteria, the evolution of these eukaryotic cells did not proceed independently of prokaryotes, but with a degree of symbiosis. Eukaryotic cells possess distinct organelles; most animal cells possess mitochondia and most plant cells contain chloroplasts. Both these organelles are approximately the same size as bacteria and possess their own DNA. Mitochondria probably originated from oxygen-utilizing bacteria, such as early examples of proteobacteria or cyanobacteria, which have been captured by eukaryotic cells to provide energy through oxidative phosphorylation. Chloroplasts probably originated from photosynthetic bacteria in order to produce energy from light. In both cases, the eukaryotic cells ingested the bacteria but did not destroy them; allowing coexistence known as endosymbiosis. Both animal and plant eukaryotic cells were taking up the energy-generating machinery of bacteria, which had evolved over millions of years, thus obviating the need for their separate evolution in eukaryotic cells. This certainly accelerated the development of eukaryotic cells. It is believed that the incorporation of mitochondria and chloroplasts into eukaryotic cells also occurred 1.5 billion years ago as eukaryotic cells emerged. The mitochondria-containing cells became the cells of animals and the chloroplast-containing cells those of plants. Both mitochondria and chloroplasts still contain their ancient DNA, which replicates independently from the nuclear DNA of the eukaryotic cell itself. Similarly, this DNA is not affected by the sexual reproduction of the host and carries unaltered DNA from generation to generation. The acquisition of these energy-producing organelles rapidly increased the evolution of the eukaryotic cells, resulting in the Cambrian explosion of multicellular, eukaryotic animals and plants 500 million years ago.

What else distinguishes a bacterial cell from a eukaryotic cell? The size is an obvious distinction: the bacterial cell could be two microns in length whereas a mammalian cell may be fifteen times longer and 1,000 times greater in volume. However, one striking feature is the maintenance of the self-replicating genetic material, DNA. In bacteria, it is maintained as a single genome, with the genes closely packed and genes controlling related functions often clustered together. On the other hand, most eukaryotic cells have their DNA divided up into several chromosomes and are diploid, receiving one set of chromosomes from each parent. Therefore the chromosomes are homologous pairs, basically carrying the same genes in the same order. This arrangement can compensate for errors in the DNA as recombination of a damaged gene with the duplicate intact gene can rectify the error. The single copy of the bacterial genome means that if a mistake is made, it is likely to be permanent and often is lethal. The reason for this may be that many eukaryotic cells form part of a much larger organism where disastrous mutations could cause major damage, such as cancer. So the duplication may have evolved to minimize this. Bacteria, on the other hand, although they are single-celled organisms, are usually not in isolation but form colonies. Each cell is capable of regenerating the colony. So if, during DNA division, mutations do occur which are lethal for the host cell, there will be enough members of the colony, in which mutations have not occurred, to keep the colony thriving.

From a technical point of view, it has been convenient to categorize bacteria according to their response to the Gram strain. This technique developed by Hans Christian Gram in 1884 distinguishes bacteria with a single membrane (Gram-positive) from those that have a double membrane and much thicker cell wall (Gram-negative). The simpler cell membrane structure of Gram-positive bacteria would suggest that the ancestral bacteria were Gram-positive; however, there is some evidence that the opposite may be true. Whichever is the case, it has proved crucial in the examination of bacteria to separate them into these two broad classes.

Chapter 2
Evolution

Bacteria and their own fight for survival

Normally bacteria divide by binary fission, whereby a single cell expands by creating macromolecules that make up the components of the cell and the cell wall. When the cell reaches approximately double its size, a septum is formed between the two halves of this enlarged cell. The septum is essentially part of the cell wall that, when complete, forms two complete daughter cells, which are able to separate from each other. In ideal conditions, this cell division can take place once every twenty minutes; so, in theory, a single bacterial cell could produce more than sixteen million progeny in eight hours. Although cell division time is short, the time taken to replicate the DNA of the cell is longer, often twice as long. In order to keep up with cell division, new replication cycles of DNA are started before the previous round has finished. Indeed, multiple copies of DNA are being replicated within a fast-growing culture. Rapid replication of DNA promotes mutations. Many of these will be lethal but some will be beneficial. In particular, the mutations have allowed individual bacterial species increased capability to compete for nutrients, often to the disadvantage of neighbouring bacteria, resulting in some local 'warfare', particularly in environments such as the soil. This led to the development of 'weapons', chemicals known as secondary metabolites that are excreted by the bacteria. Bacteriocins are released by bacteria to destroy surrounding bacteria of similar species, but they are quite limited in their effect. This capability has been refined by, for example, the Actinomyces, which are able to release chemicals that kill most bacteria; it is from these that we get most of our antibiotics.

By many small advantageous mutations, bacteria have been able to adapt to almost every environment and this has been the primary method of survival and evolution; however, the evolution of bacteria has not been without its own struggles. A hundred years ago, Frederick Twort and Félix d'Hérelle independently discovered small elements capable of killing bacteria. These were bacterial viruses or bacteriophages, simply called phages for short (Figure 2). Their existence had been implied since the earliest medical documentation which reported that some river waters could cure some bacterial infections. Recent metagenomic studies have shown that these viruses are profuse in most aqueous environments, making them as abundant as bacteria themselves.

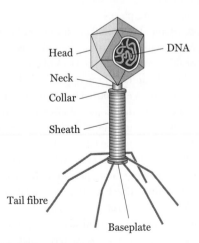

Head — DNA
Neck
Collar
Sheath
Tail fibre
Baseplate

2. **Structure of a T4 bacteriophage**

The conventional lytic phage, which is composed of DNA surrounded by a protein coat (Figure 2), follows a cycle; it starts when the phage attaches to the bacterial cell and injects its DNA. This takes over the normal replication machinery of the bacteria and results in the production of many, sometimes hundreds, of new phage protein particles. These particles pick up the replicated phage DNA; they are released, killing the host bacterial cell in the process, and are able to infect new bacterial cells. Bacteria have had to defy this attack and, quite quickly in some cases, the bacteria become resistant, often by alterations in the cell surface so that the phage cannot attach. There has been a continuous struggle between bacteria and these viruses but, as with viruses that attack humans, a balance is usually reached.

At some points in this process, there would have been an inevitable exchange of genes. The new phage particles may pick up some bacterial genes along with, or rather than, new phage DNA which they can pass on to other bacterial cells on subsequent infection. Some phages can become parasitic after infection; instead of replicating and producing new phages, the DNA integrates into the bacterial DNA. In this case, the phage DNA is replicated every time the bacterial DNA is replicated and so every daughter-cell DNA carries a copy of the phage DNA. It is often kept integrated by a repressor protein encoded by the DNA itself. This is known as a lysogenic phage. If this protein is compromised by a chemical, ultraviolet light, or some other insult, then the phage DNA initiates the production of new phage particles, the DNA replicates and is incorporated into these particles, and they are released to invade other bacteria. Often, the bacterial genes, surrounding the inserted phage DNA, are incorporated into these particles and these are spread to new bacterial cells.

Bacteria and their place in the world

Bacteria are, of course, still evolving; however, if we move out of geological time and into a time frame with which we are more familiar, the role of bacteria appears more static. They currently have an important role to play within some of the biogeochemical cycles that allow other living organisms to survive; these include the carbon, nitrogen, and sulphate cycles.

Carbon is the essential element in all living matter and it has to be converted into different forms. Carbon dioxide produced by animals and as a by-product of industry would soon render the world uninhabitable; plants fix carbon dioxide but so do bacteria. These bacteria, known as autotrophs, can produce their own organic compounds using the sun's radiation and carbon dioxide and water. They convert them to sugars by photosynthesis. These can either be used by the bacteria themselves or by other organisms. Some carbon is taken up as methane and there are methane-oxidizing bacteria that convert it to carbon dioxide. Bacteria are also primarily responsible for the removal of carbon from dead tissue and plants. On the other hand, some bacteria can use organic carbon, particularly sugars, for the production of energy with the release of carbon dioxide. So different bacterial species are on opposite sides of the cycle and the success of each will depend on the availability of the nutrients they require. If there is an increase in carbon dioxide, the photosynthetic bacteria will prosper, whereas if too much has been converted into sugar, and there is an abundance of oxygen, respiring bacteria will thrive. The burning of fossil fuels and the high quantities of carbon dioxide it produces appear to upset this balance.

Nitrogen is an important component of proteins and nucleic acids, such as DNA. Most nitrogen on Earth is available as a gaseous molecule, comprising two nitrogen atoms, in the atmosphere but this is relatively inaccessible for plants, and thus eventually ourselves. The atmospheric nitrogen has to be 'fixed' and this is mainly performed by bacteria that contain a nitrogenase enzyme that can break the bond between the nitrogen atoms. These atoms can be converted to an ammonium salt or nitrates, which are usable by plants. Indeed, the nitrogen-fixing bacteria are often found clustered around the roots of plants. Bacteria are used for most of the interconversions of the various forms of nitrogen. Some bacteria, such as *Pseudomonas* species, are even able to convert nitrite back into molecular nitrogen, which is released into the atmosphere, so equilibrium is established. This balance is upset when large quantities of nitrates are used as fertilizers and there are insufficient bacteria to convert the excess back into atmospheric molecular nitrogen, resulting in pollution.

Sulphur is an essential component of proteins and some co-factors. Plants require sulphates but this sulphur derivative is often not readily available. Sulphur is abundant as inorganic sulphides and thiosulphates, which can be oxidized by the bacterial genus *Thiobacillus* to form sulphates. These can be used directly by plants and incorporated into proteins that, in turn, can become incorporated into animal proteins. On the other hand, the balance can be maintained by anaerobic bacteria, such as the genus *Desulfovibrio,* that are able to reduce sulphates to hydrogen sulphide, which can then be oxidized to elemental sulphur.

Bacteria and the evolution of Man

The presence of bacteria must have had a major impact on the development of all species and not just because of their capability to cause disease. It is difficult to determine how this was manifested before the Cretaceous extinction sixty-five million years ago; however, when we examine modern animals, we can see how bacteria have lived in a symbiotic relationship with other organisms. Grass did not emerge until after the Cretaceous extinction and became widespread as the Earth became drier. This provided probably the greatest worldwide food resource but many animals could not access it because grass contains fibrous celluose. Many animals, such as ourselves, are simply unable to digest it. Some animals, such as cattle, evolved a system that relied upon bacteria. Their stomachs are made up of four compartments.

The grass enters the rumen, which is essentially a large fermentation chamber. There are billions of bacteria including the species *Ruminococcus flavefacians, Ruminococcus albus, Bacteriodes succinogenes, Butyrivibrio fibrisolvens*. These bacteria break up the fibre of grass cells. Cattle then regurgitate to grind the products with their teeth (chewing the cud) to try and break up the fibre further. The bacterial digestion of the grass releases nutrients that are, of course, used by the bacteria themselves but the excess is available to the animals as it is passed on to the abomasum, which contains acid similar to that in our own stomachs. The nutrients are further broken down and absorbed by the small intestine. Without bacteria, grass could not be a food source.

How have bacteria affected the evolution and development of Man? They clearly are not as crucial to the acquisition of nutrients from food as they are in ruminants, and the human gut can survive without them. When a child is born, the gastrointestinal tract is sterile but soon becomes colonized. The early colonizers are mainly aerobic bacteria such as *Escherichia coli* (*E. coli*) but quickly change during development. The gut of the infant has a low oxygen level and is soon colonized by anaerobic bacteria, similar to those found in the adult. It is estimated that up to 500 different species can colonize the gut, though probably only thirty are found in any significant numbers. The vast majority of these, more than 99 per cent, are anaerobes. Why are they in the gastrointestinal tract and are they important? They are often seen as commensal (not harmful) or even mutualistic (beneficial) bacteria and they do perform a number of important functions required for the evolution of the human population, which has learnt to live on a very varied diet. While we still do not possess enzymes to digest some carbohydrates, some bacteria do possess these enzymes and can convert them into short-chain fatty acids that we can utilize. It is estimated that we would need to ingest 30 per cent more food to maintain our body weight if we did not have bacteria fermenting in our gut. They can, of course, cause problems. Flatulence after eating baked beans is caused by the bacterial fermentation of specific bean sugars.

As a further by-product, the gut bacteria can also produce vitamins. We usually ingest these whole in our food, because we cannot make them, but some gut bacteria can produce vitamins such as vitamins K and H (biotin). The presence of large numbers of commensal bacteria in our guts also ensures that any pathogenic bacteria have a competitive disadvantage, so these bacteria go some way in protecting us against food-poisoning bacteria. The final role of these bacteria may be as stimulants to the gastric immune system, ensuring that it is primed against invasion by more pathogenic bacteria.

Moving further along the gastrointestinal tract, bile is released into the duodenum from the gall bladder. This is alkaline, which neutralizes the stomach acid, but it is bactericidal to many bacteria. There are, however, bile-tolerant bacteria, such as *Escherichia coli*, that thrive in this environment (Figure 3). The large numbers of these predominantly aerobic bacteria suppress the establishment of pathogenic bacteria; for instance,

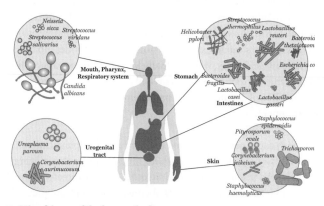

3. Microbiome of the human body

they will curb *Clostridium difficile* (colloquially known as *C. diff.*) or the bacteria causing typhoid or dysentery.

There are also significant numbers of bacteria in the mouth (Figure 3). We are aware of those that cause tooth decay under certain conditions but it is not clear whether these bacteria perform a useful task or whether they are just a remnant of an earlier role that these bacteria had had in our evolutionary development. In some reptile species, such as *Varanus komodoensis* (Komodo dragon), the mouth bacteria contain deadly bacteria (*Salmonella* species), which produce toxins, sufficient to be fatal with a single bite to their victims.

The skin is the largest organ in the body, covering an area of nearly two square metres. It provides a natural barrier against bacterial infection, which, when penetrated, can cause infection either within the skin, dermatitis, or deeper within the body. We have up to 1,000 different species of bacteria able to survive on the skin (Figure 3). Most are located on the surface and in the upper areas of the hair follicles. The surface of the skin is generally acidic and has a high salt concentration, and some bacteria, such as *Propionibacterium* and *Staphylococcus* species, thrive in this environment. As in the gut, their numbers alone discourage the growth of other, more pathogenic bacteria but these bacteria also stimulate the skin's immune responses to repel transient invasion. This response stimulates the excretion of small peptides, up to fifty amino acids long, which actually kill many species of invading pathogenic bacteria. Thus the skin and its bacteria have evolved as the first line of defence against pathogenic bacteria in the environment.

Humans could not have evolved to their current sophistication without the help of bacteria, both for their ability to obtain energy in their cells through mitochondria and also more directly in the ability to maximize the nutrients obtained from food and their primary defence against serious infection.

Chapter 3
Discovery

The signs for the existence of bacteria as infectious agents have always been present and these signs have been important in the survival of Man. The discovery of fire and the sterilization of food with cooking must have had a huge impact on the health of humans. Whether this was a conscious decision or not is impossible to determine. Certainly by the medieval era and probably as far back as Roman civilization, it was known that water may not be safe; so water purified by the presence of alcohol was frequently consumed. Wine, of course, was common in southern Europe and ale in the British Isles. There was also an understanding that some diseases were infectious, though there was no known cure; only a degree of prevention of infection could be obtained by increasing your distance from those showing symptoms. Members of the English court regularly left London when there was a plague epidemic as they were aware that plague could be transmitted from person to person. They were not aware, however, that the causative bacterium was more likely to be transmitted by the fleas of rats; rather that there was something in the air.

The discovery of bacteria came from Antonie van Leewenhoek. Van Leewenhoek was a Dutch trader living in late 17th-century Delft. Glass lenses had been known in ancient Assyria and they had been used in spectacles since the 13th century. Van Leewenhoek was able to make what are by modern standards crude lenses. The lenses were thick and essentially very powerful magnifying glasses, able to provide magnification of about 200-fold. With this crude tool and his exceptional eyesight, he was able to describe not only red blood cells but also single-cellular organisms including bacteria. He was also able to draw accurately what he saw. In 1683, he wrote to The Royal Society in London describing discoveries that he had been making since 1676, which included examining scrapings from teeth where he noticed that there were small individual cells that were moving; some were spinning and some moved rapidly through a water environment. He was also amazed by the vast numbers that were present. He called these 'animalcules'; they were probably the first bacteria ever seen. Even though microscopes improved in the 18th century, there were no further significant descriptions of bacteria until the 19th century, when compound microscopes, made of more than one lens, were able to provide clearer images. More recently the electron microscope has provided clear images of bacterial cells (Figure 4).

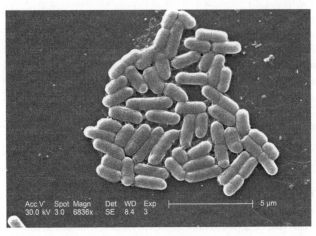

4. Electron micrograph of a Gram-negative bacterium

By the second half of the 19th century, the theories of Darwinian evolution had been accepted and that of spontaneous generation rejected, the latter being exemplified by the rejection of the view that maggots were spontaneously generated on a rotting piece of meat. However, the spontaneous generation theory still persisted for bacteria as it was thought that most food could spoil because it generated bacteria rather than becoming infected from outside. The French microbiologist Louis Pasteur had originally been a chemist and he exposed a broth full of nutrients to heat, known to kill bacteria. He showed that if this nutrient medium was in a glass flask with a long swan neck open to the air then it never became contaminated with bacteria (Figure 5); however, if the glass was broken the medium soon became contaminated and bacteria grew. From this he concluded that bacteria were present in the air and they were being introduced on particles of dust. The swan neck prevented the dust entering the vessel. Indeed, some of these flasks were kept for many years without becoming contaminated. This became the basis of Pasteur's germ theory of infection.

Pasteur examined food and beer that had spoiled. He noticed that bad beer did not contain the large round yeast cells that had been responsible for the fermentation but rather small, rod-shaped bacteria. He concluded that these were responsible for the spoilage.

5. Pasteur's swan-neck flask

Therefore, he went on to develop the technique of heating to prevent bacterial contamination of liquids by bacteria. This technique, which we now call pasteurization, raised the temperature sufficiently to ensure that the common bacteria, which were already present in these liquids and responsible for spoilage, would be killed. As this did not involve boiling, it had a lesser impact on the taste. This was not sterilization and it did not remove all bacteria but rather just prolonged the shelf life of the liquid.

Pasteur's theory that bacteria were invading from outside caught the attention of the Scottish surgeon Joseph Lister, who believed that this could be exploited to prevent the human body from being infected by bacteria. He had originally considered that miasma (bad air) was responsible for the infection of patients during surgery. Ignaz Semmelweis, a Hungarian doctor working in Vienna, had shown the importance of washing hands in the prevention of infection. Despite this, most surgeons still did not believe that this was an essential precaution. Lister read Pasteur's findings about the germ theory of infection. He tried to find a chemical that would kill bacteria and experimented with carbolic acid, which contained phenol. The dipping of instruments in carbolic acid prior to surgery radically reduced post-operation infection, and the swabbing of wounds with carbolic acid also reduced the incidence of gangrene. Lister was also adamant that his staff wash their hands between operations; he took Semmelweis's instructions further and insisted this was done using 5 per cent carbolic acid.

The germ theory of bacteria was, by the 1870s, well established and bacteria were seen as major causes of disease. Pasteur demonstrated that gangrene, an infection of dying tissue, occurred in the absence of oxygen and he was one of the first to show that there were anaerobic bacteria as well as those that require oxygen. Pasteur had also been interested in vaccines and he had found that bacteria he had cultured in the laboratory, when injected often did not cause infection in laboratory animals. Furthermore, when he did subsequently inject animals with fresh bacteria, he found that the animals were, in some way, protected. We now call these cultured bacteria 'attenuated' or 'non-virulent', and Pasteur called the process of pre-injection of attenuated bacteria for subsequent protection 'vaccination'. He used this term in honour of Edward Jenner's immunization against smallpox with the structurally related cowpox virus. Pasteur's innovation was to use the *same* bacterium for immunization as that which causes the disease.

The German scientist Robert Koch first met Pasteur in London in August 1881 following joint invitations from Lister. Their opinions on the germ theory often differed; in particular, Koch was sceptical about Pasteur's ability to attenuate bacteria to render them suitable for vaccination. He believed that the properties of bacteria were permanent and disavowed Pasteur's proposal that variations in a bacterium's virulence could explain why epidemics could occur, as the virulence of the bacterium temporarily increased. Koch's major contributions were to explore the role of individual bacteria as the sole cause of specific diseases. He developed techniques for isolating individual bacteria, particularly by the use of agar, a substance derived from seaweed that solidified media, thus producing a nutrient matrix on which individual bacterial cells could form discrete colonies. The circular glass trays which he used, Petri dishes, were named after his assistant. This was an enormous step forward and allowed the isolation of individual bacterial species. He and his team were able to isolate the various bacteria that caused diphtheria, typhoid, pneumonia, gonorrhoea, meningitis, leprosy, bubonic plague, and tetanus.

Koch's most important discoveries were the bacteria responsible for anthrax and tuberculosis. With the former, he was able to show that the anthrax bacillus was able to transfer from animal to animal. He noticed, however, that it did not survive long outside the animal and instead formed spores. This discovery contradicted his own theory on the permanent properties of bacteria for which he had argued so vehemently with Pasteur. He found spores to be metabolically inactive and that they could survive for long periods in soil, changing into active virulent bacteria on entering a new host. His techniques for bacterial isolation also identified the bacterium responsible for the tuberculosis, *Mycobacterium tuberculosis*. This was a major medical discovery as tuberculosis was responsible for more than 10 per cent of premature deaths at the time.

By the end of Koch's research career at the start of the 20th century, many of the major bacterial species responsible for infection had been identified. Although bacteria were clearly involved in the process of disease, it had not been clear whether they were solely responsible. Koch's isolation techniques allowed him to prepare pure cultures of bacteria, so he devised criteria to determine whether an individual bacterial species was the cause of an infection. These were known as Koch's Postulates. Originally they were:

1. The bacteria could be isolated from the animal suffering from the disease.
2. The bacteria could be cultured in the laboratory.
3. When the cultured bacteria were introduced in another animal, they should cause the same disease.
4. The same bacteria can be re-isolated from the second infected animal and shown to be identical to original bacteria.

If these postulates could be followed, it would be possible to state that the individual bacterium was the cause of the disease and the disease was infectious.

Koch's postulates are still used today but they have often had to be modified, particularly the first postulate, as many bacteria can just be carried by the first hosts, who become asymptomatic carriers, rather than causing an infection. An example of this may be the bacteria that cause cholera, *Vibrio cholerae*, or dysentery, *Shigella dysenteriae*. In these cases, the first postulate has to be abandoned. Furthermore, the introduction of pure bacterial cultures did not always precipitate disease in test animals, not least for the fact that the ability of a bacterium to cause disease is strongly influenced by the immune status of the new host. As we shall see later, a new generation of bacteria, now known to cause disease but only in immunocompromised patients, cannot be identified by Koch's postulates.

Causative bacteria were found for a number of different infections including *Streptococcus pneumoniae* for pneumonia by Leo Escolar in 1881 and *Haemophilus influenzae* for bronchitis by Richard Pfeiffer in 1892. The latter had been discovered during an influenza outbreak and was thought to be the causative organism of influenza. Koch's postulates could not demonstrate this and it was subsequently found that both these respiratory pathogens were actually opportunistic, in that they only caused disease in patients whose immune system was already compromised by another infection, for example by the influenza virus, or another underlying disease affecting the immune system.

Escherichia coli was one of the most important bacterial species identified at the time. Discovered in 1885, it was named after its discoverer, Theodor Escherich. Its importance has been not only because it is a pathogen, which can cause a variety of different symptoms, nor because it is one of the most important bacteria commonly found in the human gut, but rather because it has become the main scientific tool for studying bacterial structure, genetics, and virulence.

The list of bacterial discoveries is vast. We know that bacteria can survive and have been discovered in virtually every environment on Earth. They can also survive in the extremes of temperature, pressure, and acidity in the environment. The ability to survive extreme acid conditions is not confined to environmental bacteria. Gastric ulcers were suspected as the reason why Napoleon Bonaparte kept his hand within his coat, in order to relieve the pain, and their presence was confirmed by a post-mortem after his death. In his case, the ulcers had led to a malignant tumour. The cause of ulcers was suspected to result from stress or eating spicy food but it was not until 1982, when Robin Warren and Barry Marshall of the University of Western Australia identified a small, helical bacterium, that these factors were no longer associated with ulcer formation. Warren and Marshall isolated the bacterium *Helicobacter pylori*. It had previously been believed that no bacterium could live in the stomach, though there had been some pathological evidence showing these bacteria in gastric biopsies since the late 19th century. Warren tested their hypothesis by drinking a culture of *Helicobacter pylori*. Within five days, he developed the symptoms of gastritis, which he eventually eradicated with a course of antibiotics. This was a hugely important discovery not only because it identified a new species of bacteria but also because ulcers were finally able to be treated not just with histamine H_2-receptor antagonists (such as ranitidine), the conventional treatment to relieve the symptoms, but also with antibiotics to remove the cause. Because of its predisposition to cause ulcers and the subsequent risk that this damage may lead to malignancy, *Helicobacter pylori* has been identified as a possible carcinogen which should be eradicated.

Other, previously unidentified, bacteria were identified in the final quarter of the 20th century. Closely related to *Helicobacter pylori* are the bacteria of the genus *Campylobacter*. Suspicions that an organism such as this existed were originally expressed by Escherich in 1886 where he saw spiral-shaped bacteria in stools of children with diarrhoea. Similar observations were made until this organism was first isolated in the 1972 by Dekeyser and Butzler, who examined the faeces of a young woman who had severe symptoms of diarrhoea and fever. This prompted a search to find this organism in the faeces of other patients with severe diarrhoea.

This organism was found to be responsible in the cases of about 5 per cent of children with severe diarrhoeal symptoms. Two species were subsequently identified, *Campylobacter jejunii* and *Campylobacter coli*. *Campylobacter jejunii* was subsequently found to have the ability to become invasive (rarely it could spread through the body and was not just confined to infection of the gastrointestinal tract). By the 1980s, *Campylobacter jejunii* infections had become well recognized and were shown to be the major cause of bacterial enterocolitis.

Gastrointestinal bacteria have not been the only recent bacterial identifications. In July 1976, there was a convention of the American Legion at the Bellevue Hotel in Philadelphia. They had gathered to celebrate the bicentenary of the signing of the American Declaration of Independence. A large number of the veterans subsequently fell ill with pneumonia symptoms; over 200 were given medical treatment and thirty-four died from this unknown disease. These were reported first to the Pennsylvania Department of Health and the outbreak was then referred to the Center for Diseases Control and Prevention (CDC) in Atlanta. They considered that the environment of the hotel might be the cause rather than human-to-human spread as there was no subsequent infection away from the hotel. Furthermore, some pedestrians walking past the hotel also became ill. Still no cause could be found. The lungs of patients who had died of the mysterious disease were examined and nothing could be found except some rod-shaped bacteria that were ignored at the time. With all subsequent investigations proving fruitless, Joseph McDade at CDC decided to re-examine these rod-shaped bacteria. He produced a serological test for this organism and found that the blood of all the sufferers had antibodies to this bacterium. This was a previously completely unknown organism and was named *Legionella pneumophila*. It was living in the water-cooling part of the hotel's air conditioning system and was thus spread throughout the hotel. The disease has been called Legionnaire's Disease.

Legionnaire's Disease is an example of environmental bacteria that have previously had no major contact with Man, which is probably why we were unaware of them. However, when we alter the environment in which we live, we can change it sufficiently to allow previously harmless bacteria to prosper. *Legionella pneumophila* is a widely distributed bacterium that lives in water all over the world, especially in temperatures above 25°C. It is completely harmless unless it has access to lungs. This did not happen until we sprayed or atomized water. Thus Legionnaire's Disease is solely a product of modern living. Of course, many regulations have been introduced to reduce its impact but outbreaks still occur and approximately 500 people become infected every year in England and Wales.

The isolation of new bacteria as a result of changing lifestyles is becoming more common. There has been increasing concern about *Escherichia coli* or *E. coli*. As we have seen, these bacteria are regular residents of our gastrointestinal tract. However, there are many types of this organism, some of which lead to some serious symptoms. It is the most common cause of traveller's diarrhoea, but this usually resolves within a couple of days. Our vulnerability to this is just that our immune systems have not experienced a particular localized strain of this bacterium before, whereas the local population have become resistant to it. More serious is the emergence of new strains of *E. coli*, which were previously unknown. The most notorious of these is *E. coli* O157:H7. The number refers to it serotype and it is often abbreviated to *E. coli* O157. Its importance lies in its capability to cause very severe haemorrhagic diarrhoea, which can result in kidney failure. Rarely these bacteria can be spread from person to person but usually outbreaks result from contaminated meat or vegetables. It was first isolated in a single patient in 1975 but it was first shown to cause an outbreak in 1982, caused by the ingestion of contaminated hamburgers. Most gastrointestinal bacteria are spread by the faecal–oral route, and *E. coli* O157:H7 is no exception except that the bacteria originate predominantly from the faeces of cattle. Without causing the animal any symptoms, *E. coli* O157:H7 reside in the rectum of cattle, probably within the majority of herds in the United Kingdom. The reason why they should reside there now and did not in the past is unclear but is probably related to changes in animal husbandry. We know that cattle have, since the Second World War, been fed a variety of experimental diets; the Mad Cow Disease (BSE) outbreak was a result of this. Something we have changed appears to have influenced the microenvironment within the rectum. Cattle continue to shed these bacteria in the faeces and the meat obtained from them becomes contaminated. Only a few *E. coli* O157:H7 strain types appear to be virulent to humans. In these cases, if the meat is contaminated, while it is uncooked it can pass the bacteria to other food products particularly during preparation. Interestingly meat that has not been minced is less likely to cause problems as the outer sections of the meat are the first to get cooked and this is usually sufficient to kill the bacteria. On the other hand, if it is minced, the contaminated outside of the meat is ground up and will now be in the middle, the area that is the last to be heated during cooking.

There are still probably many more bacteria that have yet to be discovered. If there are more bacteria in a kilogram of soil than there are humans on the planet, it is highly likely that we are currently unaware of the majority of bacteria species that exist. It is estimated that we know fewer than 5 per cent of all the bacteria present. Because of their direct involvement with us, the bacteria that we are most knowledgeable about are those that infect us or influence our lives. Furthermore, the speed at which bacteria are able to evolve implies that we shall never know the full extent of this, the most prevalent biomass on the planet.

Chapter 4
Environment and civilization

Bacterial spoilage of food

From the rise of prehistoric Man to the end of the 19th century, and in some areas to the present day, our main involvement with bacteria had been as a result of their ability to spoil food and liquids. The paradox of collective hunting and agriculture to provide for the tribe was that the food gathered could not be stored, particularly in warm and humid climates. This was particularly acute with animal products, which were already contaminated with bacteria poised to damage it. It must have been clear early on that food which had spoiled was dangerous to health. To some extent all animals had the excessive acidity of the stomach as the primary barrier against bacterial invasion.

It is not known whether the cooking of food was developed as a means of sterilizing it or whether it was considered to improve the taste. Whichever it was, this has a significant impact in reducing bacterial contamination. The ability to reduce the bacterial load was important for human health; however, even exposing food to heat to kill the bacteria was only a temporary solution, for once the heat source had been removed the food easily became contaminated again. A longer-term solution was required. Probably the earliest method of preserving food was to 'salt' it. This was achieved by removing as much water as possible and then soaking the food in high salt concentrations. The salt draws water out of many bacteria by osmotic pressure and largely prevents them growing. The value of salt as a preservative has been known for many millennia and salt became a highly prized commodity. Indeed, the word salary derives from the Latin for salt, which was given to Roman soldiers for the preservation of their food.

Another ancient method of preserving food was to smoke it by exposing it to the smoke of the fire. This probably originated from the use of smoke to deter insects from landing on meat, and the observation that the meat took longer to spoil. The smoke contains particulates that enter the meat and inhibit bacteria growth. Unlike salt, this is often considered to be a flavour enhancer and is still continued in the preparation of meats such as pastrami and fish such as salmon.

Honey as a non-spoiling foodstuff has been known for 10,000 years as its collection has been depicted on early cave paintings. It was taken from wild bees, as it still is in more primitive cultures. Honey also preserves other food. It has a very low water content and its high sugar concentration can inhibit fermentation. Its viscosity also provides a barrier preventing bacteria from penetrating and this, coupled with its sticky nature, provides a good defence against bacteria invasion either for itself or any foodstuff that is placed within it. Many honeys contain antibacterial compounds that actively kill bacteria; one is glucose oxidase, which converts glucose into gluconic acid with the release of hydrogen peroxide gas. This gas is very effective at killing bacteria. Manuka honey from New Zealand is very effective in killing methicillin-resistant *Staphylococcus aureus* (MRSA); it can also produce hydrogen peroxide, but there are other antibacterials present in it as well.

Alcohol production, which is not normally a bacterial process, was probably discovered about 7,000 years ago, the first evidence of its production going back to ancient Persia. In sufficient concentration it can also act as a food preservative. How long this has been known is difficult to judge. The Romans were probably the first to distil alcohol but it is not known if they used it as a preservative. Robert Boyle in the 17th century identified it as a preservative and it has been used for preserving food ever since.

Bacteria of the *Acetobacter* genus can oxidize alcohol, converting it to acetic acid (vinegar). These bacteria can proliferate in up to 7 per cent alcohol. Vinegar was known to the ancient Egyptians and Babylonians and probably was first identified as it destroyed wine. Because *Acetobacter* requires oxygen, wine kept in sealed containers is not normally affected. Vinegar was found to provide a sharp taste and has been added to many foodstuffs at a concentration of up to 10 per cent. It is also a preservative and can be used for pickling but usually at a much higher concentration (about 20 per cent). Almost any natural alcohol product can be made into vinegar including beer, wine, and cider. The important aspect is to ensure that just the right oxygen level is maintained as too much will remove all the carbohydrates present. The ability of vinegar to maintain its characteristics over a long period are seen with the making of classic balsamic vinegar, where the vinegar is stored for a minimum of 12 years, though some is kept far longer.

Cheese manufacturing

Hunter-gatherers probably became 'farmers' in the Mesolithic period after the last ice age. The domestication of animals, probably first sheep and goats, probably took place 9,000 years ago in the Middle East, with herds of cattle being kept within 1,000 years afterwards. An early result of domestication would have been the ready provision of milk from each of these animals. Milk is highly nutritious and a good energy source, but it spoils easily and is soon unfit to drink, especially in hot climates. Some means of preservation was required.

The addition of acid to milk with rennet causes the milk protein casein to coagulate and thus solidify. The mixture of solids and liquids, known as curds and whey, is separated and the solid curds are compressed into a block. This block of solidified milk, which we call cheese, is able to resist spoiling by bacteria because they will have difficulty in penetrating any further than the outer layer and the block remains acidic, which can be too hostile an environment for bacteria to become established. Therefore, the production of cheese preserves milk, often for many weeks, and it conserves most of the fat, protein, and minerals that were in the original milk.

It is unlikely that early cheese production was a conscious effort to acidify milk but rather a chance observation that it could happen spontaneously if the milk became infected with certain bacteria, which convert the sugar lactose in milk into lactic acid. The likely explanation has been that nomads in the Middle East often placed milk into a sheep or goat stomach for storage during a journey and when they came to drink it, it had coagulated because of the bacteria in the stomach and the residual rennet present. Whatever the story, basic cheese production probably occurred soon after the domestication of sheep and goats. We have no means of knowing what early bacteria produced the first cheeses but they were probably closely related to those that are currently employed. These are collectively known as lactic acid bacteria and a number of bacteria can achieve this, but the most common are *Lactococcus lactis* subsp. *lactis* or *cremoris*, *Lactococcus salivarius* subsp. *thermophilus*, *Lactobacillus delbruckii* subsp. *bulgaricus*, and *Lactobacillus helveticus*. The genus of *Lactococcus* had previously been identified as *Streptococcus* and often these bacteria are identified in this genus.

The first recorded observations of cheese production were in Egyptian tomb murals, probably from around 4,000 years ago, and in Babylon at about the same time. Ancient Greek mythology, including Homer, mentions cheese production and by the Roman civilization, cheese was a normal part of the diet. It is likely that all of these were basic cheeses, made with a single starter bacterial culture comprising a bacterium related to the ones above (Figure 6). The results would have been moist salty cheeses such as modern-day sheep's cheeses, for example feta from Greece or beyaz peynir from Turkey.

In the past 2,000 years, cheese manufacture has become more sophisticated, with specific bacteria used for individual cheese types. The bacteria were naturally present and the effect exploited.

lecoxta. oplo. fri a huia. Cecerio q fir æ puip lacre.uina. nutrir i inpinguar. Ꭺ ceuni.

6. Cheese production in the 14th century

The so-called mesophilic *Lactococcus lactis* are often used in the production of the European hard cheeses, such as cheddar, feta, and the Dutch cheeses. These bacteria do not ferment citrate and do not produce carbon dioxide. The softer cheeses are often formed with the thermophilic bacteria such as *Lactobacillus delbruckii* subsp. *bulgaricus* and *Lactobacillus helveticus*. These produce cheeses such as mozzarella and Emmental. Of course, the range of cheeses is now vast and there is much experimentation with new strains of bacteria and even mixtures of mesophilic and thermophilic bacteria.

In addition to the initial challenge with the starter bacterial culture, it was soon found that the introduction of other microorganisms enhanced flavour and texture. Much of this has been achieved by the use of environmental bacteria that were found serendipitously during the cheese-making process long before the existence of bacteria was actually known. The involvement of the fungus *Penicillium notatum*, which gives blue cheese its distinctive colour, is well known but the involvement of secondary bacterial cultures is less familiar. These are known as adjunct cultures. Recognizable products of an adjunct culture are the 'eyes' or 'holes' in Swiss cheese. The starter bacterium is likely to be *Lactococcus lactis* followed by the late addition of the adjunct bacterium, such as *Propionibacter shermani* or *Propionibacterium freudenreichii*. These bacteria use the lactic acid produced by the starter bacteria and slowly generate carbon dioxide. This gas produces the 'eyes' of Emmental.

More often adjunct bacterial cultures are used to alter texture and taste. The raw cheese is often salty and bitter, from the lactic acid, so the use of adjunct cultures can change the final taste by breaking down some of the products of the primary culture. They also often speed up cheese ripening. Bacterial adjunct cultures are usually employed in the manufacture of harder cheeses as the softer ones are usually treated with moulds, not bacteria. They are often used for washing the outside of cheese as a smear, giving it the characteristic outer surface. These smear bacteria have to be salt tolerant and their ability to grow at low pH is the main reason why they were originally found on the surface of cheese.

Lactobacillus casei and *Lactobacillus plantarum* are now added during the manufacture of cheddar. *Brevibacterium linens* has been used to produce the outer surface of Gruyère, Munster, and Limburger cheeses. This bacterium produces the very characteristic red-orange colour of the cheese and is also responsible for the distinctive flavour. Historically, cheese manufacture was often done in damp dairies where many of these bacteria would have already existed. Now, of course, these bacteria cultures are carefully controlled and the correct mixture of adjunct bacteria is added to provide a more uniform product. Finally and equally important, these smears protect against moulds spoiling the cheese and allow the cheese to be stored for months and sometimes years. The principle for this is that a large mass of bacteria prevents the establishment of other microorganisms that might spoil the food. This is probably achieved by the bacteria releasing chemicals that prevent the growth of other bacteria rather than direct competition for nutrients. In any case, the net result is that these 'good' bacteria prevent 'bad' bacteria from spoiling the food and hence are a potent form of food preservation.

Yogurt manufacture

Yogurt is a Turkish word and means to thicken or coagulate. Yogurt has probably been used for more than 6,000 years, though the first written reports were during the Roman empire where it had been noted that nomadic groups were preserving milk by thickening it and that the product tasted pleasant. Its origins are probably similar to cheese production in that the milk had been kept in sacks made of animal products where it had coagulated. When the sack had been emptied and refilled with milk, the coagulation would have occurred even more quickly. Towards the end of his reign in the mid 16th century, Francis I of France had formed an alliance with the Turkish Ottoman empire. He developed a severe bout of diarrhoea, which his physicians could not cure. A Turkish physician gave him yogurt, which cured him, and its benefits as a medicine were then widely broadcast.

Because of the Turkish occupation of central Europe during the 16th and 17th centuries, it became a part of the staple diet of many of the conquered countries. Its popularity extended throughout Asia, from Sumatra to Russia, both as a food such as raita and as a drink such as lassi, both from India.

By the 19th century, it was noted that Bulgarians often lived longer than their Western counterparts. The Bulgarian microbiologist Stamen Grigov examined Bulgarian yogurt microscopically and noted that there were large numbers of both rod-shaped and spherical bacteria. The rod-shaped bacteria we now call *Lactobacillus delbrueckii* subsp. *Bulgaricus*, though at the time it was just referred to as *Lactobacillus bulgaricus*. The associated longevity of regular consumers of yogurt resulted in its promotion as a health food, which is where it stands today.

Yogurt is traditionally made by adding a starter culture, often called a 'mother culture', which is usually taken from a recent batch of yogurt, and adding it to milk at about 40°C (Figure 7). Nowadays the milk is sterilized by heating it to near boiling first to kill any resident bacteria. In the traditional production of yogurt, the milk was not sterilized and the high concentration of bacteria in the starter culture usually, but not always, suppressed the resident bacteria. The original production of yogurt must have been a fortuitous event, for it requires both the spherical bacteria seen by Grigov, *Streptococcus thermophilus*, as well as the *Lactobacillus*. Both bacteria are thermophilic and prefer elevated temperatures, above 40°C for growth, a temperature that is lethal to many other bacteria. This probably explains why yogurt originally developed in southern Asia. The culture in yogurt

Milk is pasteurised at 85°C for 30 minutes or 95°C for 10 minutes

Milk is homogenised to improve yogurt consistency

Milk is cooled to 40–45°C and inoculated with a starter culture of *Lactobacillus bulgaricus* and *Streptococcus thermophilus*

Mixture is incubated at this temperature for several hours (usually until a pH of 4.5 is obtained), while the bacteria digest milk proteins and ferment lactose to lactic acid

The yoghurt thickens and is cooled to 5°C to stop fermentation

The yoghurt thickens and is cooled to 5°C to stop fermentation

Flavourings and fruit may be added before packaging

7. Basic flow chart of modern yogurt production

produces acid; hence the sharp taste if it is not sweetened. However, it is this acidity that preserves it and prevents other bacteria and moulds from spoiling it. Like cheese, it will now keep for many weeks even if not kept cold.

Probiotics

Yogurt is essentially the first probiotic medium. A probiotic is, according to the Food and Agriculture Organization of the United Nations, 'a live microorganism which when administered in adequate amounts confers a health benefit on the host'. Probiotic was a term first employed by Werner Kollath while working on dietary problems at the University of Munich in 1953, when he described beneficial bacteria in digestion in contrast to those that cause disease.

The presence of lactic acid-producing bacteria in milk is not a fortuitous accident. They have co-evolved as they help to protect young mammals, fed initially on breast milk, against gastrointestinal diseases. This is achieved by providing significant quantities of bacteria, in breast milk, which are able to colonize the gastrointestinal tract and create an acidic environment, so that more pathogenic bacteria cannot survive. As well as lactic acid-producing bacteria, human milk contains bifidobacteria as well. There are many species of the *Bifidobacterium* genus and they can colonize many parts of the human body. They are anaerobic bacteria so they can only operate effectively where there are low oxygen levels, similar to conditions in the lower gut. Bifidobacteria have the ability to ferment oligosaccharides; these are relatively short carbohydrates made up of chains of up to about ten sugars (oligosaccharides). In particular, they can ferment the oligosaccharides found in milk. This would certainly assist breastfed infants. Some strains of bifidobacteria can ferment oligosaccharides derived from plants, which would be difficult to metabolize otherwise and thus their presence could aid absorption of nutrients from plant material. The common species used as a probiotic is *Bifidobacterium bifidum*; it is a natural colonizer of the gut and it does maintain the balance of microorganisms, preventing the growth of other bacteria. It is also believed to strengthen the immune system but much of the research in this area is still ongoing and the benefits have yet to be proven.

Probably the main benefit associated with probiotic bacteria is recolonization of the gut after antibiotic treatment. The effect of some antibiotics can be effectively to annihilate all the main residents of the lower bowel, leaving it vulnerable to infection by pathogenic bacteria; an example of this will be seen later with *Clostridum difficile*. The introduction of large numbers of beneficial bacteria can restore the balance within the gut before it is eventually recolonized by other harmless bacteria.

Sewage disposal

In London before the 19th century, there had always been an underlying stench resulting from the disposal of human waste, mainly in communal cesspits. It is the reason why the affluent areas of the city moved westward as the prevailing westerly wind carried the smell towards the east of the city. The summer of 1848 was unusually hot, the cesspits overflowed, and raw sewage flowed into the Thames. The hot weather encouraged bacteria to grow and this produced an overwhelming stench, known as the 'Great Stink'. This affected business in the Houses of Parliament and, although heavy rain soon dealt with the immediate problem, it was the precursor to the building of London's sewerage system. However, unsatisfactory open sewerage arrangements still exist in much of the developing world.

Bacteria have always been an essential component of the disposal of human waste. It appears to be the natural instinct of almost all mammals to defecate away from the immediate living area and often to use a communal area. This could certainly be found in Roman civilization where collective latrines collected human waste, which was often disposed of in a cesspit. The collected waste is decomposed most quickly by anaerobic bacteria, which originate in the faecal matter itself. Eventually these bacteria would decompose all the matter; however, in a cesspit there is no drain and, as it is continuously filled, eventually it has to be emptied or abandoned. Many human communities still use cesspits, though their modification in the industrialized world is the septic tank, so called because of its use of bacteria to break down human waste (Figure 8). This is usually a sealed tank, favouring anaerobic conditions and preventing unwelcome odours. Unlike the cesspit, there is a soakaway where the liquids from the bacteria fermentation are able to drain into the surrounding soil. So a balance is reached such that the new waste coming in is offset against the discarded bacterial 'waste', as long as the system does not become clogged or the bacteria are not killed by the disposal of detergents or other chemicals.

A sewage treatment plant is essentially a septic tank on a much larger scale. Human waste is transported by means of water through sewers to a treatment plant, which may be many

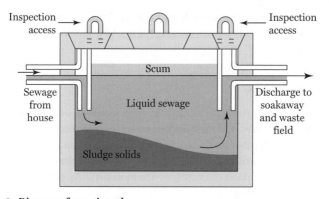

8. Diagram of a septic tank

kilometres away. Sewage treatment plants often add bacteria after the non-human items, such as fats, are removed. There are many different bacteria found in sewage treatment plants, other than the normal intestinal organisms; they include mainly soil bacteria such as those of the *Bacillus*, *Pseudomonas*, and *Achromobacter* genera. In essence, bacteria detoxify human waste, whether this is a natural process or in dedicated treatment centres.

Animal decomposition

Bacteria are also responsible for a considerable part of body decomposition after death. In the natural state, where an animal dies in the open, the rate of decomposition is often accelerated by maggot activity; however, bacterial activity has usually preceded this. After the heart stops, the oxygen in the tissues is quickly depleted by the aerobic bacteria present, leaving a largely oxygen-depleted environment in which the anaerobic bacteria from the gut and respiratory tract can now proliferate. The animal's own cells break down and the released proteins, fats, and carbohydrates allow these bacteria to produce acid and gas, often methane and hydrogen sulphide. This gives the bloated appearance following death and eventually these gases are released, which is responsible for the smell. This is known as putrefaction. This process slows as the available nutrients are depleted.

Composting

The death of plants also involves bacteria, usually those that are found in the soil. The most obvious example is the decomposition of leaves after they have fallen from deciduous trees in the autumn. Bacteria are not alone in achieving this decomposition as fungi are also major contributors. The most useful by-product of this is composting.

The Romans realized that if horticultural waste was piled up, it had generally decomposed by the following growing season and could be used as a fertilizer for the soil. It was a labour-saving method not only to dispose of waste but also to recycle it. This composting is not a natural process as most decay takes place in thin layers on the surface, not in large piles. So it often had to be 'seeded' artificially, usually with some soil. In this soil, there is a range of bacteria which, with the help of some fungi, normally oxidize the plant matter. In their natural state, these bacteria have ready access to oxygen, which they require, as they are on the surface. During composting, the centre of the pile is likely to become oxygen deficient and thus the pile needs to be mixed to ensure adequate oxygenation. Many of the bacteria are thermophilic and the temperature starts to rise: the reason why steam is often seen coming off some compost heaps. However, if this is not controlled, then heat may kill the bacteria causing the oxidation. Water is also required to provide a sufficiently aqueous environment but the active bacteria are aerobic and too much water will prevent access of air and consequently oxygen. The net result is that the bacteria break down the cellulose and cell structure, releasing carbon and nitrogen, along with phosphorous and potassium that is then used as fertilizer in further cultivation.

Chapter 5
Bacterial pathogenesis

Most people, when they consider bacteria, identify them as a cause of disease. It is relevant to ask why bacteria became pathogenic to Man. The reason is straightforward: in order to provide them with a source of nutrition and as a means of spread. Actually very virulent bacteria, which cause death quickly, are often not in the optimum situation for their spread. Although not toxic itself, the soil bacterium *Clostridium botulinum* produces toxins, which can be produced if this bacterium is in tinned food that has been insufficiently sterilized. In the anaerobic conditions in the tin, the toxins are produced. They are amongst the deadliest known. Ingestion of these botulinum neurotoxins gives an almost immediate reaction as there is reduced acetylcholine release, which blocks the electrical transmission to the nerves controlling the muscles. The paralysis caused soon affects breathing, leading to death. The incapacity of the patient drastically reduces the ability of the organism to spread and the bacteria remain trapped in the dying patient. A similar situation is found with bubonic plague, caused by *Yesinia pestis*: unless the bacterium spreads to the respiratory tract and becomes airborne, which is rare, then the bacterium dies with the patient. So hyper-pathogenic is not necessarily beneficial.

Stages of pathogenesis

It is likely that bacteria were not originally pathogenic and that they have developed this trait through evolutionary changes. We are almost certainly not the first animal that most bacteria infected. This would have first been through invertebrates and then eventually through mammals and on to us. The transmission of an infectious organism from animals to cause disease in humans is known as zoonosis and it is considered that most bacterial infections started this way, though some bacteria may now have evolved further so that they infect only Man.

The successful invading pathogenic bacteria have to establish themselves at the optimum site for nutrition and subsequent release from its host. They are equipped with molecules, known as virulence factors, which allow them to achieve this and they broadly fall into four groups; adhesins, aggressins, impedins, and invasins (Figure 9). The names are almost self-explanatory. The adhesins allow the bacteria to bind preferentially at specific positions to ensure that the bacteria are optimally placed.

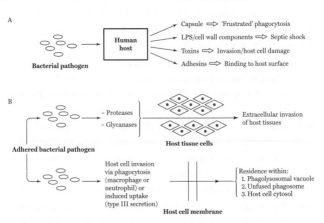

9. Diagram of some of the pathogenicity factors associated with bacteria. In A, the bacteria evade the bodies defences. In B, once it has adhered it may further invade the body

An example of these are the fimbriae that position bacteria responsible for diarrhoea high in the gastrointestinal tract, away from the normal commensal bacteria of the lower bowel, giving them an advantage as they have earlier access to the nutrients passing down the gut. The aggressins are usually enzymes that have a localized effect, often producing a barrier between the bacteria and the host defences. The common pathogen *Staphylococcus aureus* produces coagulase, converting fibrin from fibrinogen and effectively placing the invading bacteria within a clot, preventing access to the host defences. Impedins act against the individual defence systems themselves. They include the ability to produce a protective capsule, such as that which surrounds *Streptococcus pneumoniae*, without which it cannot shield itself from the defences in the lungs. They even include molecules that can degrade antibodies such as the two cysteine proteinases in *Streptococcus pyogenes*. The integrity of the antibody molecule is held together by cysteine-cysteine disulphide bonds which these enzymes destroy. Invasins allow the bacteria to move within the host once the initial infection has been established. *E. coli* O157:H7 is a pathogen that differs from the commensal *E. coli* in that it is able to invade the host by migrating out of the gastrointestinal tract into the bloodstream.

Whatever the host, in the first instance the bacterium has to enter. In humans, there are many entry points. The most obvious are through ingestion in the gastrointestinal tract, through inhalation into the respiratory tract, and via contact with the skin. The skin is a natural barrier to bacteria and protects against infection; however, if it is broken either through a small insect bite or a larger wound, bacterial invasion can soon follow. There are other routes as well, such as through the eyes, the sexual organs, and the nose. Some infections are caused by commensal bacteria already within the body that have, for some reason such as an underlying disease, been able to grow and spread. This is usually termed an opportunistic infection, though some opportunistic pathogens may have originated outside the body.

Once inside, the bacterium has to establish itself quickly. Many bacteria just start growing in the epithelial cells close to the point of entry. This is characterized by the boils caused by *Staphylococcus aureus* on the skin or the establishment of gonorrhoea by *Neisseria gonorrhoeae* on the mucus membranes of the sexual organs. In the lungs and intestine, there are macrophages, which are non-specific phagocytic cells of the host immune system, that migrate quickly to the incoming bacteria, engulf, and destroy them (Figure 10). This is usually the first line of host defence. After engulfing the bacteria, the macrophages return to the lymph nodes with the key surface structures of the invading pathogen, thus stimulating acquired specific immunity against future invasion.

The triggering of macrophages stimulates other immune responses. Some bacteria can cross the epithelial cells to cause systemic infection; *Mycobacterium tuberculosis* crosses the epithelium in the lungs to cause pulmonary tuberculosis, *Salmonella enterica* serovar Typhi crosses the epithelial of the intestinal tract as a precursor to causing typhoid. So the bacteria are able to leave the localized environment at the entry point and become systemic, entering the bloodstream or the lymphatic system. When some bacteria pass through the epithelium, they

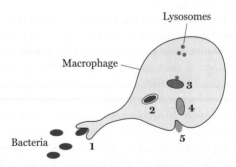

10. **Phagocytosis of bacterial cells by a macrophage. Bacteria are engulfed by the macrophage (1), then surrounded by a phagosome (2). A lysosome fuses with phagosome (3) and the lysosomal enzymes break down the bacterium (4) which is then excreted from the macrophage (5)**

are held up by the basement membrane, which can act as a barrier in itself.

As the bacteria reach this point, other body defences start. Complement is a compound within the blood and other bodily fluids that kills bacteria on contact, but this is often circumnavigated. However, the migration of bacteria identifies their presence and the host produces chemotaxins and cytokines. These identify the position of the pathogen and the local phagocytes migrate towards them, following the track of the chemotaxins. The defence system also releases opsonins, which bind to the invaders and mark them for the phagocytes, greatly increasing the chances of recognition and destruction. It used to be thought that the trigger for this response was the lipopolysaccharide at the cell surface but more recently it has been found that proteins, carbohydrates, and lipids can all stimulate cytokines in mammalian cells. These are collectively known as modulins.

This is the start of the inflammatory response, the result of which gives us the feeling of burning at the site of infection. Indeed, much of the pain associated with infection is often due not to the invading bacterium but rather to our body's response to it. Successful bacterial infection, however, has to overcome the body's defences and many bacteria have the ability to do so: for example, streptococci produce streptolysins, which attack membranes and release the contents of the host cell, whether this is a red blood cell or a phagocytic defence cell; and the β-toxin of *Clostridium perfringens* causes the necrotizing of endothelial cells that make them leaky and destroy their integrity. The main role of these bacterial compounds is to interfere with or annihilate the host defence system and thus they often influence the activity of cytokines, macrophages, and neutrophils and may cause localized host cell death, or apoptosis. The net effect of which is to prevent the host defences reaching the invading bacteria.

As the infection proceeds, the bacteria will usually replicate and this often generates a range of responses from the immune system. The body temperature may increase to slow bacterial growth and, as long as it does not go too high, is a potent method of control. This is often coupled with a feeling of malaise. If the body cannot cope at this stage, either a chronic disease establishes itself or the infection may even lead to death. Usually, though, the transport of the key bacterial components by the macrophages to the lymph nodes initiates the formation, within a couple of weeks, of specific antibodies which specifically seek out the individual bacteria that caused the infection. It would explain why bronchitis, often caused by *Haemophilus influenzae*, takes about three weeks to resolve if not treated quickly with antibiotics. This is known as acquired immunity and is the basis of vaccination.

Once the bacteria have been able to multiply, they need to escape to infect other hosts. Death of the host often inhibits this release of bacteria, so death itself is not usually directed by the bacteria; more often it is due to tissue damage from the host defences. Escape is important and bacteria produce many mechanisms to achieve this. The most obvious are the toxins. The enterotoxin of *Staphylococcus aureus* is released when the organism is growing, often on food outside the body. When ingested, it elicits violent vomiting, thus spreading the bacteria to the environment. Similarly, the diarrhoea-causing bacteria often produce a toxin that induces rapid and frequent bowel movements, to release the bacteria back to the environment. The most extreme example is *Vibrio cholerae*, which causes such severe diarrhoea that this alone can cause death through rapid dehydration if untreated. It also demonstrates how we have controlled infections without the use of antibiotics but rather through improved hygiene. In developed countries, human waste is collected and treated, so the route for the spread of bacteria such as *Vibrio cholerae* has been broken. The cough induced by *Mycobacterium tuberculosis* in TB is the mechanism to promote its spread; however, changes in lifestyle in the developed world have considerably reduced the risk of infection because individual dwellings are now less crowded and fewer people share the same room. Diptheria, caused by *Corynebacterium diptheriae*, was often spread via aerosol droplets to many members of the same family, especially if they slept in the same room. The sexually transmitted bacteria are released in the same manner as they were acquired, two mucous membranes coming into close contact, one of which was infected.

Toxins

Many of the symptoms that we associate with bacterial infection actually result from the toxins bacteria excrete; indeed the symptoms may develop even if there are no live bacteria, provided the toxin has been released. A simple initial demarcation is to classify them as endotoxins or exotoxins.

Endotoxins are usually part of the bacterial cell, often found on the surface (Figure 11). They are not released until the bacteria are destroyed by the immune system. When released, endotoxins can induce inflammation but can also lead to septic shock if there are too many endotoxins in the blood. Many endotoxins are thought to derive from the lipopolysaccharides that make up the cell surface. Although antibodies are effective against lipopolysaccharides, they are more active against intact molecules; the endotoxins are usually smaller components. They are responsible for increases in body temperature and the lowering of blood pressure. The bacterium *Neisseria meningitidis*, the cause of rapid onset of and severe meningitis in young adults, has a waxy coat that contains an endotoxin. It is the level of endotoxin that makes this organism so lethal; it is often more than 100-fold greater than other bacterial types. Although better known for invading the meninges, it is particularly deadly if it can produce sepsis in the blood. The endotoxin can rupture blood vessels, leading to severe haemorrhaging. The basic treatment for *Neisseria meningitides* infection is aggressive antibiotic therapy; however, one of the products of antibiotic treatment is that the bacterial cell is broken up and this releases endotoxin. So certain antibiotics can actually exacerbate the clinical manifestation of an infection rather than resolve it, if an endotoxin might be present.

Exotoxins are molecules that are released from live bacterial cells. They are usually the component that causes the clinical manifestation of diseases and it is not always necessary for the bacterium to be present (Figure 11). The enterotoxin of *Staphylococcus aureus* and the botulinun toxin of *Clostridium botulinum*, described previously, are clear examples. The neurotoxic properties of botulinun toxin A have been put to cosmetic and medical use. Botox has been used to remove wrinkles because it causes flaccid paralysis of local muscles, thus removing wrinkles that arise from muscle contraction. Its effect on wrinkles is not permanent and wears off after a few months. Less well publicized is its medical use, where it was first used to treat eye disorders such as continuous blinking or eyes that were not properly aligned with each other, such as cross-eyed. It worked in the same way by relaxing the muscles. However, its use has become far more extensive, for example, to control sweating, migraines, and incontinence, amongst many other conditions. It is perhaps surprising that such a powerful poison can be put to so many different uses. With its frequent cosmetic use, it is easy to forget how deadly botulinum toxin can actually be. One gram is thought to be sufficient to kill one million people. Such concentrations are never normally reached and the death rate from ingestion of botulinum toxin without suitable treatment is around 60 per cent. This is usually due to respiratory failure and

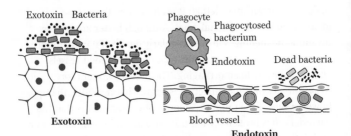

11. General comparison of endotoxin and exotoxin action

the rate can be radically reduced if the patient is given antitoxin and put on an artificial respirator. Symptoms occur after about twelve hours.

Other exotoxins are less beneficial and can be as deadly. *Bacillus anthracis* produces a three-protein exotoxin. One component, called protective antigen, binds the host cell and there are two enzymes known as oedema factor and lethal factor. These three proteins individually have little effect but they form a complex. The protective antigen transports the two enzymes into the cell. Once inside, the oedema factor increases the level of cyclic AMP (cAMP), which controls the water balance inside the cell. Increasing the level of cAMP increases water retention, hence the oedema, and this lessens the effect of the host defence macrophages. The lethal factor destroys the proteins of the cell, leading to cell death. This is an extremely effective mechanism and patients who contract an anthrax infection soon become very ill. Anthrax can be contracted through inhalation or direct contact. Although anthrax is relatively rare, through the skin is currently the most common method to contract the infection and, if the bacterium invades by this method, it is rarely fatal. Gastrointestinal infection from infected food is very rare and can cause severe diarrhoea. Severe anthrax infection usually results from inhaling spores of the bacteria into the lungs and its mortality rate is greater than 90 per cent, with death occuring within two days after the onset of symptoms. About 10,000 inhaled spores are sufficient to cause death. Traditionally, those who worked with animals, particularly their hides, were vulnerable to infection. It was called 'wool-sorter's disease' for this reason. The bacterium relies on the death of the host to spread further. In the dead carcass, it forms spores, which then may become airborne or ingested by other animals.

One of the most deadly exotoxins known is produced by *Clostridium tetani*, the organism that causes tetanus. When the organism enters damaged tissue, it produces a neurotoxin which is inactive while still inside the living bacterium and is not released until the bacterium dies. Once activated by proteases, the toxin amplifies the signals from nerves to the muscles and this can give the characteristic spasms associated with tetanus. Although often identified in the head, hence the name 'lockjaw', actually muscles all over the body are affected, including those for breathing. Death often follows even with the use of artificial respiration. In unvaccinated and untreated patients, death is almost inevitable.

The bacterium that has had the greatest impact on the European population over the last two millenia is *Yersinia pestis*, the cause of plague. Though little was known about it at the time, it is thought to have been responsible for reducing the British population from four to one million in the middle of the 6th century. Initial transmission to humans is from a bite from the infected fleas of rats. The bacterium blocks the proventriculus of the flea, effectively causing a state of starvation in the flea, which becomes desperate to feed. On finding a victim, the flea tries to feed from the blood but vomits the content of its stomach into the bite wound and thus into the victim's bloodstream. Once inside the new host, the bacteria accumulate in the lymph nodes and they multiply extraordinarily quickly, causing the lymph nodes to swell considerably; these are known as buboes and are the distinctive symptom of infection. The bacterium releases a toxin complex (Tc) and virulence factors. These have the effect of inhibiting the phagocytes, preventing the neutrophils accumulating, promoting cell apoptosis (death), and the breakdown of clots, thus promoting the release of the bacteria into the bloodstream. If the bacteria get into the lungs before death, they can multiply quickly as they will prevent the body's defences here also. While bubonic plague is a self-limiting disease, if the bacteria proliferate in the lungs, then it becomes pneumonic, producing fever and a productive cough with bloody or watery sputum. These form drops which may be inhaled by those in close face-to-face contact with the patient. This form of plague is very virulent and was responsible for the death of a third of the English population in the epidemic of 1348–9, known as the Black Death. It had originated in China and was carried along the Silk Road and by ship.

Virulence factors encoded by 'foreign' DNA

The ability to be virulent, particularly in humans, is likely to be a later development in the evolution of bacteria and results from their adaptation to this new environment. It takes a long time for bacteria to evolve individual genes and this can be achieved much more quickly on DNA that is not part of the bacterial chromosome. The location of virulence genes on extrachromosomal DNA ensures, first, that if mutations occur in the evolution of the genes they are not lethal to the bacterium as they might be if they occurred in the bacterium's own DNA; and second, as these pieces of DNA are mobile, genes that have evolved in one bacterium can be passed easily to others. Some genes known to cause virulence are located on bacteriophages, and without these genes the bacteria would not cause an infection. These phages have, at some time in the past, integrated themselves into the bacterium's DNA. They may remain dormant, known as a prophage, and eventually be induced to release new phages. On the other hand, the bacterium may go through what is known as lysogenic conversion whereby the genes in the prophage start to express conferring novel trait(s) on the bacterial cell. The diphtheria toxin produced by *Corynebacterium diptheriae* and primarily responsible for the disease is located on such a lysogenic bacteriophage. Similarly, the toxins responsible for cholera produced by *Vibrio cholerae*, botulism produced by *Clostridium botulinum*, and scarlet fever produced by *Streptococcus pyogenes* are all derived from lysogenic phages.

Plasmids and pathogenicity islands

Plasmids, another form of extrachromosomal DNA found in bacteria, are also known to encode virulence factors, and this is common in some species. The virulence of non-typhoid *Salmonella* species is often encoded on plasmids. In particular, the *spv* gene of *Salmonella typhimurium*, which allows the pathogen to invade the liver and spleen, is located on a transferable plasmid. The product of this gene considerably reduces the number of bacteria required to cause a severe infection.

Plasmids in *E. coli* can carry genes that allow the bacteria to colonize the small intestine, where bacterial numbers are usually very low. This allows the bacteria to have earlier access to the nutrients passing down the gastrointestinal tract, removing the need to compete with the bacteria of the large intestine. The gene responsible for the attachment produces fimbriae, which allow the bacteria to attach to enterocyte cells. Alone, this attachment does not cause severe symptoms but is usually accompanied by the presence of two plasmid-encoded enterotoxins, heat stable (ST) and heat labile (LT) toxins. Both toxins stimulate loss of electrolytes and water excretion from the human enterocytes; hence the symptoms of diarrhoea. Both the attachment fimbriae and the toxins are required for maximum effect. These *E. coli* are known as enterotoxigenic, or ETECs, and they are the common cause of 'traveller's diarrhoea', notably because visitors do not have the antibodies to these local bacteria. These bacteria are not invasive and the symptoms are usually self-limiting as long as the patient remains well hydrated. The plasmid-encoded fimbriae in uropathogenic *E. coli* (UPEC), operate in a similar manner, with the adhesion allowing the formation of a micro-colony in the urethra or the bladder. In this case, it is purely the position of the colonization that is important and no toxins are involved. They produce the classic symptoms of a urinary tract infection. These are rare in healthy men, rather more common in women as the urethra is considerably shorter and most infections are caused by the introduction of bacteria through the urethra. Often self-limiting and usually easily treatable with antibiotics, the infection can become more severe if the bacteria are able to travel up into the kidneys, where the establishment of an infection can cause pylonephritis.

Enteroaggregative *E. coli* (EAEC) also produce watery diarrhoea like ETECs. A separate plasmid produces aggregative adhesion fimbriae that aggregate human cells, stacking them like bricks. The bacteria also produce a haemolysin and a stable toxin but the pathogenic mechanisms are less well understood. Infections caused by EAECs, currently far less common than those caused by ETECs, are increasing particularly in North America.

The *E. coli* that make the media headlines are usually enteropathogenic (EPEC) or enterohaemorrhagic (EHEC) *E. coli*. These bacteria do not use fimbriae for attachment. Instead, the bacteria come into close proximity with the epithelial cells of the gut and form a characteristic attachment to these cells and localized destruction (effacing) of the microvilli. The 'attaching and effacing' genes are located in the bacterial chromosome in an area known as locus of enterocyte effacement or LEE region. The *eae* gene in this region encodes intimin, an adhesin similar to one responsible for cellular invasion in *Yersinia pestis*. This forms the close attachment. The LEE region also encodes the *tir* gene. The Tir protein is transported to the human cell where it acts as a receptor for intimin; in other words, the bacterium produces not only the means by which the bacterium binds to the human cell but also exports the mechanism by which the human cell accepts this binding. EHECs cause more severe infections than EPECs as they produce a phage-encoded cytotoxin, similar to that found in *Shigella*, known as verotoxin. This toxin induces bloody diarrhoea and can directly attack the kidney cells. Platelet numbers are severely reduced by clotting (thrombocytopenia) leading to haemolytic anaemia, known as haemolytic-uremic syndrome (HUS). The most infamous EHEC strain, *E. coli* O157:H7, has been responsible for a number of outbreaks and many deaths but there are other strains as well, including strains O145 and O104; strain O145:H4 was responsible for the major outbreak in Germany in 2011.

Pathogenicity islands

The LEE region of these pathogenic *E. coli* is located on the bacterial chromosome, but the region itself is known as a pathogenicity island (PI). PIs can contain clusters of virulence genes, which can include adherence genes, toxins, invasins, etc. They are called islands because they can usually be defined as DNA that is distinct from the rest of the bacterial chromosome and has thus been imported into the bacterium by some form of gene transfer; perhaps originally on plasmids, phage, or transposons, which may have long since been lost. The tell-tale signs, however, are usually present; the region is usually flanked by direct repeated sequences of DNA. The sequences at either end are identical, and they are associated with transposase and integrase genes, which would promote movement between chromosome and mobile DNA. They are often closely associated with transfer RNA genes. The composition of their DNA is different from the rest of the bacterial chromosome, indicating that their origins are different. The movement of discrete blocks of DNA, carrying virulence genes, from one bacterium to another means that the evolutionary solution that once caused one strain of bacteria to develop these genes may be passed on directly to other strains, without them having to be 'relearned'. Besides *E. coli*, PIs are found in *Salmonella enterica*, *Helicobacter pylori*, *Shigella* species, *Yersinia pestis*, and *Vibrio cholerae*. The similarity of the attachment and effacing genes in *Helicobacter*, the invasive nature of *Salmonella* and *Yersinia* and toxins in *Shigella* and *Vibrio* with those found in various pathogenic strains of *E. coli* suggest that, although they are not identical, they probably had a common source. In these cases, the transfer of the virulence genes has accelerated the ability of non-pathogenic species to become pathogenic. Many of these strains of bacteria are pathogenic only to humans and there has been insufficient time since the emergence of Man for these pathogenicity mechanisms to have evolved independently.

Flesh-eating bacteria

Until twenty years ago, most people had not heard of flesh-eating bacteria, or rather necrotizing fasciitis. It is an extremely rare infection in previously healthy individuals but rather more common in patients with some other underlying disease or a medical procedure that has compromised the immune system. This is probably the reason we hear more of it now. Because of its dramatic and rapid effects, it also attracts the attention of the press and hence it acquired the label of flesh-eating bacteria. A number of different bacteria can cause this type of infection but the most common is *Streptococcus pyogenes*, but only very specialized strains with specific virulence determinants. These are actually very rare. The bacterium does not 'eat' but rather causes an infection of the skin and particularly the subcutaneous layers. It is the speed at which it infects and the very rapid spread of the infection through the cutaneous tissues that has earned this infection its fearsome reputation. Infection starts at a point of injury, which may be severe, such as an operation incision, or even undetected, which allows the bacterium access. Inflammation occurs if the infection is near the surface and this may soon turn to necrosis or death of the tissue. Early and aggressive intravenous antibiotic therapy can limit the damage of the infection, though the dead tissue has to be removed. However, most doctors have never seen a case of necrotizing fasciitis and may not recognize it, particularly if the initial infection is deep. It is extremely serious if untreated and will often lead to death.

Vector-borne diseases

These are normally associated with viruses and parasites. The devastating example in the past has been plague but nowadays the tick-borne bacterium from the genus *Borrelia* causes more medical problems. It was first identified in the town of Lyme, Connecticut in 1975, where the causative bacterium was *Borrelia burgdorferi*. The tick often ingests the bacterium after biting deer; so the disease is often found in areas frequented by deer. If the tick subsequently bites another animal, as with the case of plague-bearing fleas it vomits the bacterium into the bloodstream. If not treated early with antibiotics, the eventual symptoms can become disabling and lead to arthritis and paraplegia. It has spread to areas of the United Kingdom and Europe where there are deer; in this case the causative bacteria are different species, *Borrelia afzelii* and *Borrelia garinii*.

Tooth decay and gum disease

We associate bacteria with identifiable infections of the body but every day we take precautions to limit the damage caused by the bacteria in our mouth. We brush our teeth to prevent the tooth decay (dental caries) and gum diseases (periodontal disease) caused by the bacteria in our mouths. There are many different species of oral bacteria but the most damaging are often those of the *Streptococcus* genus. Examination of skulls from people who had lived up to the 16th century shows a relatively low incidence of damage due to tooth decay. Then, the major sweetener had been honey, which contains antibiotic properties that would prevent its use as a nutrition source for oral bacteria. The colonization of the West Indies and the importation of sugar (sucrose), particularly as it became available as an everyday commodity, soon led to widespread and serious tooth decay. Sucrose provides a nutrition source for the bacteria, which they ferment. The result of this fermentation, which is similar to what we have seen in the production of yogurt, is the production of acid. It is this acid that destroys the enamel of the tooth and causes cavities. Tooth decay began to be controlled first by the introduction of bristle toothbrushes by William Addis in the early 19th century, although their use did not become widespread until the latter part of the century. They were first used with abrasive tooth powders and, from the start of the 20th century, toothpastes, which were originally mixtures of hydrogen peroxide (an antibacterial) and baking soda.

Tooth decay was still a serious problem by the middle of the 20th century. It had been known that fluoride prevented tooth decay, by replacing hydroxide ions in the tooth enamel, making it much more resistant to acid attack. Initially there was controversy about the use of fluoride but these fears were allayed when the remarkable effects of fluoride in preventing cavities were recognized and the risks were deemed to be small. Now fluoride is regularly added to both toothpaste and water supplies, with a consequent reduction in tooth decay.

Pandemics

It is probably not appreciated that bacterial infections cause considerably more deaths than those caused by viruses; not least for the reason that the actual cause of death associated with a virus infection is often manifested by a secondary infection caused by a bacterium. The virus infection debilitates the immune system and the bacterium causes an acute infection that is often responsible for death. Pandemics are now usually associated with virus infections, whether these are swine flu, SARS, or AIDS. This was not always the case.

Plague

The most infamous bacterial pandemic in the British Isles occurred in the summer of 1348. It is known as the second plague pandemic. The Black Death, caused by *Yersinia pestis*, originated in Asia and was first brought to the British Isles by black rats travelling on a ship that landed in the west of England. It soon spread during the next year to the rest of the British Isles. Over the next two years, plague killed about a third of the whole population of the country and half the population of London. The reasons for its rapid spread are not hard to understand; at the time families tended to be large and they all lived and slept largely in one room. Hygiene was poor and rats came into close contact. Rat fleas did not normally bite humans but if infected, as described earlier, they would jump onto any available mammalian host. It is impossible to imagine what the loss of a third of the population would mean in current society but its effects were devastating in the 14th century. Prior to the Black Death, the population had largely remained in their towns and villages, but by the start of the 1350s there were severe labour shortages and the peasants started to travel to exploit this and demand higher wages. This lead to inflation, and the Statute of Labourers was introduced in 1351 prohibiting the movement of labourers and fixing wages at their pre-Black Death levels. The restrictions that this imposed led eventually to the Peasants' Revolt of 1381. This demonstrates that even in the 14th century, the loss of such a huge proportion of the population led to the undermining of society.

Plague returned to the British Isles regularly until the middle of the 17th century when improved living conditions and implementation of measures to prevent spread by the pneumonic form (Figure 12) lessened the capacity of the population to become infected and the ability of the bacterium to spread. However, it was found in sporadic outbreaks around the world, such as those in Marseille and Moscow in the 18th century, but largely petered out. The modern or third pandemic again started in China in the mid 19th century. It was exported by ship from Hong Kong to Bombay and killed more than twelve million people in India and China. It spread across the world to all continents and was introduced into the United States through the port of San Francisco. By now, the causative bacillus had been isolated and the method of transmission by rat fleas discovered, enabling stringent controls to be introduced that prevented the carriage of rats on ships. In the United States, the last major outbreak was in the city of Los Angeles in the mid 1920s. The advent of antibiotics ensured that any cases could be rapidly treated and the spread of the

Bacterial pathogenesis

12. Engraving showing the masks used to prevent contracting plague in 1665

organism was quickly controlled. The pandemic dwindled and now only a few cases annually, from around fifteen countries, are reported. The residue of the pandemic remains in the wild rodent population of the Sierras in California; though this is rarely transmitted to humans.

Cholera

A bacterium that is equally swift at killing is *Vibrio cholerae*, the causative bacterium for cholera. This water-borne bacterium has also been responsible for pandemics, seven in the last 200 years according to the World Health Organization. Because of the low speed of transport and the rapid onset of symptoms, before the mid 19th century, cholera outbreaks were largely contained in and around the country in which they emerged. The first pandemic originated in the Indian subcontinent or China, in common with those pandemics that followed. Cholera was endemic in the region surrounding the Ganges, and pilgrims to the festivals carried the organism back to their own villages and towns. This had happened in previous outbreaks but, by the 1820s, British soldiers carried it to areas far from the subcontinent including Africa and the rest of Asia.

The second pandemic spread more quickly during the 1830s as communications had improved. Again starting in India, it spread to Europe, Russia, Canada, and the United States. It was responsible for a number of deaths in the British Isles up to the end of the 1840s. However, the cause of the infection was not understood until the third pandemic reached London in 1853 and John Snow identified that contamination of the water supply was responsible. Even so, this pandemic was responsible for more than 10,000 deaths in London; however, it was the last time that cholera seriously affected the United Kingdom. Further pandemics occurred during the 19th century but better hygiene largely prevented cholera seriously affecting Europe, except for the fifth pandemic, causing a major outbreak in Hamburg in 1892, and some parts of Eastern Europe affected by the sixth pandemic in the early 20th century.

We are still experiencing the final and seventh pandemic. It started in the 1960s, this time in Indonesia. The causative bacterium was not the classic strain but a related strain called El Tor. It had originally been isolated from a cholera patient being treated in El Tor in Egypt in 1905. The strain spread from Indonesia to the subcontinent and North Africa. From North Africa it spread to Italy, though the number of cases was relatively few. This strain spread to South America, in particular Peru and northern Chile, and from there to the southern United States. El Tor is a milder pathogen than the classic strain and its symptoms are less severe; however, the main controls of the current pandemic have been better hygiene, preventing infection in the first place, and much more effective treatments, reducing mortality.

Leprosy and tuberculosis

Plague and cholera are infections of rapid onset and swift mortality. It is thus easy to follow their spread across the world. Two related bacteria are far slower in their transmission; they are *Mycobacterium leprae* and *Mycobacterium tuberculosis*, the causative bacteria for leprosy and tuberculosis. For this reason they have been much more difficult to control. Leprosy and plague were considered to be the two major epidemics of the Middle Ages and leprosy remained so until the 18th century. Certainly leprosy was common in Europe, though it did not follow classic pandemic distribution. Indeed, in many areas it was probably endemic. Leprosy is an infection of the skin and nerves; although it is contagious, a person has to have long exposure before the bacterium can establish an infection. There has been a progressive decrease in the number of infections, partly due to better and more rapid diagnosis and subsequent antibiotic treatment but also possibly to less crowded living conditions and an improved host immune status. India still has the majority of cases, mostly in poor communities, but the disease is found in other countries of the subcontinent, parts of Africa, and in Brazil. Currently, the proportion of cases in the world is now only 10 per cent compared with the past so this is a bacterial pandemic that is also declining.

The same used to be true for tuberculosis as it was for most of the major bacterial infections (Figure 13) but now the number of cases is increasing. Also slow growing, *Mycobacterium tuberculosis* can spread more quickly as it is usually airborne and coughing is a usual symptom. The universality of air travel is thought to promote the spread of tuberculosis but there is no strong evidence for this. Probably the major cause of the resurgence of tuberculosis is the increase in infections by Human Immunodeficiency Virus (HIV), which undermines the body's defences against tuberculosis so that it is a common accompanying infection. Tuberculosis is a disease particularly of the urban poor, not just in the developing world but also in the

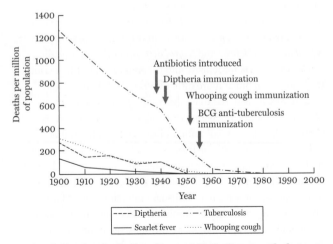

13. Graph showing the decline of bacterial infections over the last century

ghettos of industrialized countries, with inadequate access to healthcare being a further contributing factor.

Infection can only occur from those suffering from active infection and is transmitted on droplets. The infective dose of *Mycobacterium tuberculosis* is as low as 10 bacterial cells, so any droplet, in theory, could initiate an infection. Prolonged contact increases the risk, as does the virulence of the strain and how well the environment is ventilated. The vast majority of those infected do not develop the disease but it may remain latent within those infected. The risk of progression to active tuberculosis infection depends on immune status, due to either infection or previous history of transplant requiring immunosuppression, cancer and other underlying medical conditions. Low body mass is also associated with active infection, which probably accounts for its increased transmission in deprived communities and, partly for these reasons, the epidemic has been difficult to control. Furthermore, a major risk factor is diabetes mellitus, and the number of these cases is currently increasing worldwide.

The extent of the tuberculosis epidemic at the start of the 20th century was a major incentive to find drugs that could cure it. However, the 'White Death' was beginning to decline in the United Kingdom. Living conditions improved and at some point around 1935, the number of rooms in a family dwelling exceeded the number within the family. This had a considerable influence in reducing the spread of all types of infection, not just tuberculosis (Figure 13). It was, however, the discovery of streptomycin, one of the earliest antibiotics, that broke the tuberculosis epidemic cycle in the United Kingdom, and appropriate timely treatment can provide a cure (Figure 14). In many areas, poor medical supervision and insufficient dosing have resulted in a resurgence of antibiotic-resistant tuberculosis. Treatment of tuberculosis requires daily administration of antibiotics for as long as six months, almost impossible in some

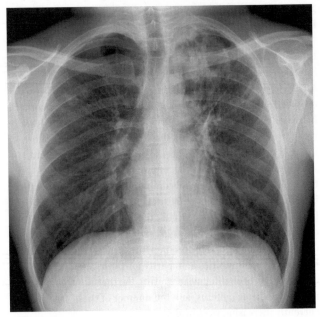

14. A chest X-ray of a person with pulmonary tuberculosis

countries. There are very few options for infections caused by these bacteria. One method of control has been vaccination with the Bacillus Calmette-Guérin, commonly known as BCG, an attenuated live tuberculosis strain from cattle. It does provide some protection for about fifteen years but largely in those countries where tuberculosis is on the decline. The efficacy of BCG is considerably lower in countries where the bacterium is still responsible for an epidemic.

The extent of the epidemic has ensured that one-third of the world's population has been infected. Though most are asymptomatic, in some countries the proportion of the population with active infections is around 2 per cent and it is compounded by the fact that 20 per cent of these strains are resistant to the main antibiotics. Until this situation is reversed, there is little prospect that the epidemic will dissipate.

Sexually transmitted infections

Currently, the most feared sexually transmitted disease is caused by a virus, Human Immunodeiciency Virus (HIV). Nearly 1 per cent of the world's population is infected. In the United Kingdom, the figure is under 0.2 per cent but rises to 5 per cent in sub-Saharan Africa. In contrast, it has been suggested that at the turn of the 20th century, 10 per cent of the adult male population in the United Kingdom had syphilis. This bacterial epidemic, then untreatable with antibiotics, was as shocking as the HIV epidemic is today and had an eventual mortality rate of up to 50 per cent. The social taboos of the time led many men to have their early sexual experiences with women who had had many sexual partners, particularly prostitutes, who provided a reservoir for the epidemic. The bacterium, *Treponema pallidum*, is commonly believed to have to come to Europe in the 15th century from the Americas, though there is some evidence that it was prevalent in Ancient Greece and Rome. It spread rapidly during the Renaissance, facilitated by the migrations occurring at that time and, because of its symptoms producing pustules like smallpox, soon earned its nickname of 'the Pox'. Syphilis was widely treated with mercury but it was probably due the promotion of condoms, particularly by the military in the late 19th and early 20th centuries, and the introduction of Salvarsan in 1912 that the incidence began to decrease. The advent of antibiotics reduced the epidemic so that by the end of the 20th century there were only sporadic cases in the developed world, though it is still prevalent outside Europe and North America.

Other major bacterial sexually transmitted diseases include chlamydia and gonorrhea, and these have not shown such a marked decline; they are still the two most common sexually transmitted bacterial infections. Gonorrhea, caused by the bacterium *Neisseria gonorrhoeae*, has been documented in England since at least the 12th century and has been presumptively diagnosed in men by a burning sensation while urinating. The problem with gonorrhea is that a woman has a greater risk of becoming infected from an infected man than a man from an infected woman. As it is initially asymptomatic in many women, the infection can spread unnoticed, making it difficult to track. The work of the clinics that treat infected patients is not just the administration of antibiotic but also tracing previous sexual partners. Through this extraordinarily difficult and sensitive task, many contacts can be treated who have shown no symptoms. The incidence of gonorrhea fell with the introduction of condoms, antibiotics, and more recently fears about infection with HIV. Early treatment of gonorrhea soon clears the infection but, if left, it can cause serious medical problems.

Chlamydia is a more prevalent sexually transmitted disease. Like gonorrhea, most women show no symptoms when infected and the bacterium can remain undetected for years. In both sexes it can lead to inflammation in the genitals, but this often occurs long after the bacterium has been passed to other partners. The bacterium, *Chlamydia trachomatis*, only multiplies in human cells and can be difficult to detect. Indeed, it cannot be detected by culture but rather by modern DNA techniques. These factors ensure that chlamydia is a silent pandemic as many cases are not detected and current assessments of the infection rate are almost certainly underestimates.

Bacterial toxins as weapons

In the Middle Ages, during the siege of a town as the population was starving, contaminated meat or even faeces were often catapulted over the wall in an attempt to weaken the besieged population by disease. There were even attempts to introduce plague into these towns. The problem with this early form of germ warfare is that the aggressors could not control the extent of the infections and were often infected themselves once the town had been taken.

A bacterium was needed that could be contained and did not pose a threat to the army using it. Anthrax, caused by *Bacillus anthracis*, was a convenient choice. This spore-forming bacterium has a very high mortality rate, does not spread extensively from its target area and does not spread from person to person. The bacillus was isolated by Robert Koch in 1876 and attempts were made to use it during the First World War, particularly against livestock and the horses of the cavalry on both sides. This was not very effective. So during the Second World War, Gruniard Island off the west coast of Scotland was 'bombed' with a particularly virulent strain of anthrax bacilli in 1942. Sheep on the island soon succumbed to anthrax. The conclusions drawn were that an anthrax bomb would provide a potent weapon against German cities; however, it had not been possible to decontaminate the island and it was concluded that bombed areas would be uninhabitable for years. In order to determine how long the strain would persist, sheep were introduced onto the island annually and they always became infected. In 1986, attempts were made to decontaminate the island with formaldehyde and much of the topsoil was removed. There have been no cases of anthrax since then.

The failure to decontaminate Gruniard Island meant that anthrax was not deployed during the Second World War. The threat of anthrax as a biological weapon has emerged periodically since then when individuals have tried to send anthrax spores through the post, but these are largely unsuccessful, though five people were killed in the attacks of 2001. Other biologicals have been developed, not least the botulinum toxin by Iraq in 1991. Sufficient was produced to annihilate the population of the world by inhalation but the associated problem of controlling the damage to the target could never be overcome. Furthermore, this toxin soon deteriorates in the presence of air. Most nations have now signed the Biological and Toxic Weapons Convention banning the use and development of infective agents as weapons. The risk of these biological weapons of mass destruction is likely always to be present but the prerequisite for containment may inhibit their use.

Chapter 6
Antibiotics

It is likely that a few antimicrobials have been used by some indigenous populations for centuries. Quinine, from the bark of the cinchona tree, had been used by Indians in Peru, and was effective against malaria; this knowledge was brought back to Europe in the 17th century when it became possible to control malaria. However, there are more than a hundred prescription medicines that are derived from the plants of the rainforest, and some have antibacterial properties. As the indigenous populations became adept at exploiting these plants, they were able to control some bacterial infections.

Oliver Wendell Holmes published a paper in the *New England Quarterly Journal of Medicine* on the 'The Contagiousness of Puerperal Fever' in 1843. At the time, the role of bacteria, or any microorganism, as a cause of infection was unknown. Holmes concluded that puerperal fever, contracted by women during childbirth, came from person-to-person contact probably from the physicians treating them. He also noted that a number of physicians had died after performing autopsies on patients who had been infected. Holmes suggested that for physicians who had performed autopsies on patients who had died of puerperal fever, their surgical instruments should be sterilized and their clothing should be burnt. His most drastic conclusion was that the physician should not participate in childbirth for the following six months.

Ignaz Semmelweis, a Hungarian doctor working in Vienna in 1847, had a less drastic solution. He noticed that in those wards where the doctors washed their hands after they had performed autopsies on puerperal fever patients, there was a much lower incidence of puerperal fever than in the wards where they did not wash their hands. He instigated a policy that all doctors performing autopsies should wash their hands in dilute calcium hypochlorite. This had the effect of reducing the cases of puerperal fever by 90 per cent. This is probably the first documented use of chemicals to control infection, albeit not on the patients themselves.

Florence Nightingale was sent to Scutari in 1854 during the Crimean War. At the time, deaths amongst the soldiers from severe bacterial infections including dysentery and cholera exceeded deaths from the battlefield by ten times. She used her experiences here to campaign for better sanitary conditions within hospitals and to recommend changes in hospital design to reduce cross-infection. She understood that the patients were infecting each other. Improvements in hygiene, which prevent infection, are known as asepsis.

Only seven years later, the American Civil War changed warfare. It was the first mechanized war and the casualties were far greater and more severe than in previous combats. The extent of injuries attracted infection and this often meant amputation. George Tichernor, a Confederate doctor, experimented with the use of alcohol on wounds. In 1863, he was wounded himself in the leg and instead of allowing its amputation, he insisted that it was treated with alcohol; the wound healed. He successfully treated many wounds. This was the first use of chemicals to control infection on the patient and Tichernor went on after the war to patent his alcohol 'antiseptic'.

In 1865, Joseph Lister, a surgeon from Glasgow, noticed like Semmelweis that there were fewer infections in births administered by midwives than by surgeons and he attributed this to poor hand-washing by the latter. Surgeons were instructed to wear gloves during all surgery and to wash their hands in carbolic acid, a phenol-based solution. Lister also found that spraying swabs with carbolic acid considerably reduced the incidence of gangrene and other infections. This is considered to be the start of concerted antisepsis, the control of infections with chemicals.

These chemicals were extremely effective but they were not sufficiently selective to be administered inside the body; rather they had to be applied to infections on the body's surface. At the end of the 19th century, bacterial infections were the primary cause of premature death in the developed world. The high incidence of diseases such as syphilis and tuberculosis drove the search for chemicals that could selectively cure a bacterial infection. It had been Pasteur who had amalgamated the germ theory of infection and the role of bacteria. He and others had noticed that some patients who had gastrointestinal infections improved if they ate blue cheese, the blue streaks in which were caused by *Penicillium notatum*. He also showed that bacillus causing anthrax would not grow if it had first been contaminated with some airborne moulds and he recognized that this might have potential in therapy; however, Pasteur thought that vaccines were the only efficient method to control infection, and lost interest.

By now bacteria were firmly recognized as the cause of many infections. Paul Ehrlich knew that some dyes were able to stain bacteria while leaving other cells unstained. He argued that this showed that some chemicals bind specifically to bacteria but not to human cells and, therefore, this could be the basis of a chemical-based therapy that selectively killed bacteria and could be administered internally to the patient. He was the first to coin the term 'magic bullet' to describe this concept and he started a search for such a solution. Starting with nitrogenous dyes, he chemically altered them little by little. When he obtained a new chemical, he tested it against suitable bacteria. He created arsphenamine in 1909, which was marketed by Hoechst AG, just one year later, as Salvarsan, for the treatment of syphilis. It was a huge improvement on the administration of inorganic mercury, which was being used at the time and caused unpleasant side effects. Arsphenamine is an arsenic-based compound but it was only partially selective; its severe side effects sometimes led to death. Modification of the molecule to give the more soluble Neosalvarsan reduced the side effects and it remained the primary treatment for syphilis for the next thirty-five years.

Military casualties in the First World War were on an unprecedented scale. Witnessing them had a profound effect on a German medical orderly and a British doctor, Gerhard Domagk and Alexander Fleming. They both noticed that most deaths came from infections in wounds rather than direct battle casualties and that some medicine was required that could be given systemically to combat bacterial infection. After the war, Domagk studied medicine and Fleming started to experiment with the use of lysozyme, an enzyme that attacks the cell wall of bacteria, in an attempt to use it as an antibacterial drug. Although effective on bacteria in Petri dishes, lysozyme is a large protein and would not survive in the body, excluding it as a viable therapy option for systemic use.

During his experiments, Fleming made the iconic serendipitous observation that colonies of *Staphylococcus aureus* on nutrient agar in a Petri dish were killed if there was a specific fungus also growing on the agar (Figure 15). The fungus was *Penicillium notatum* and Fleming realized that, as the fungus could kill the bacteria at some distance, it must be releasing a chemical. Indeed, he surmised that the chemical must have a small molecular weight if it was able to travel so far through the relatively viscous agar. He named this unknown compound penicillin. Fleming purified the *Penicillium notatum* and repeated his initial observation but this time substituting other pathogenic bacteria; it was able to kill almost all Gram-positive bacteria but did not have a significant

15. Sir Alexander Fleming in his laboratory at St Mary's Hospital, London

effect on Gram-negative bacteria. He prepared large cultures of *Penicillium notatum*, removed the fungus and kept the broth. It was an active antibacterial agent. Even when it was diluted, he found it had significant inhibitory properties. The destruction of one organism by another had already been designated as antibiosis by Paul Vuillemin in 1889, so Fleming called penicillin an antibiotic.

There are reports dating back to the 19th century where a microorganism could kill a bacterial culture, Lister had shown that *Penicillium glaucum* eradicated bacteria and Pasteur had demonstrated that patients treated with non-pathogenic bacteria became immune to anthrax; only Fleming realized that these microorganisms were releasing a compound that could be used as systemic therapy. He did not succeed in purifying the compound and it proved to be unstable. He published his findings in 1929 and they were then ignored for ten years.

A few years after medical qualification, Domagk started working in the pathology laboratories of I. G. Farbenindustrie (now Bayer AG) in 1927. He was interested in the bacterium *Streptococcus*. Domagk had been influenced by Ehrlich but thought that Ehrlich's tests were not sufficiently rigorous. He devised an animal model in mice to examine the effects of compounds being manufactured by I. G. Farbenindustrie on animals infected with streptococci. He argued that an antibacterial compound only needed to prevent the bacteria growing and the body defences would do the rest. In 1932, Domagk tested a red dye, later called Prontosil. Mice treated with Prontosil all survived infection with *Streptococcus* whereas untreated mice died. Like Fleming, Domagk found that Prontosil was active against Gram-positive bacteria but not against Gram-negative; however, it was absorbed when taken orally and it was readily excreted through the kidneys, which meant that it did not accumulate.

Prontosil was tested for two years in patients with infections caused by Gram-positive bacteria, with remarkable results, curing previously fatal illnesses. In the Pasteur Institute in Paris, it was found that the active component of Prontosil was sulphanilamide, a sulphonamide. So pure sulphanilamide was used as therapy instead and a search was made for other sulphonamides which were able to inhibit bacteria that were unaffected by sulphanilamide. In particular, the search was for a sulphonamide that would cure tuberculosis. Domagk's research slowed with the outbreak of war and, despite opposition from army surgeons, he was able to demonstrate that sulphonamides were crucial in preventing gangrene in soldiers wounded in battle.

In contrast, the war stimulated interest in antibacterials amongst the Allies. In the late 1930s, Ernst Chain was researching the use of chemicals that killed bacteria in the University of Oxford and he read Fleming's paper on penicillin published many years earlier. A culture of *Penicillium notatum* had been given to the university some years before and Chain and his boss Howard Florey cultured it but also found that the penicillin produced was hard to purify. Incorrectly assuming it was an enzyme, they freeze-dried it, which not only allowed significant concentration but also stabilized it. As chemists, they were also able to remove any impurities and produced a product one million times more active than Fleming's extract. In 1941, after extensive animal experiments, they tried the purified penicillin on a policeman dying of septicaemia caused by *Staphylococcus aureus*. He was treated every three hours with penicillin, his temperature dropped, and he improved until the supply of penicillin was exhausted, when he worsened and eventually died. Clinical trials were continued once sufficient quantities had been produced and miraculous cures were achieved. Laboratories in the USA were recruited into production, which were producing large quantities in the last year or so of the war. Many allied soldiers will have owed their lives to penicillin but it was still only effective against Gram-positive bacteria and many of the wounds had been infected by Gram-negative bacteria, especially if they had become contaminated with soil. There was no cure for these and often penicillin eradicated the Gram-positive bacteria only to allow Gram-negative infections to take over.

The discovery of penicillin ensured that many microbiologists searched for new antibiotic-producing organisms, particularly from the soil. Selman Waksman at Rutgers University in New Jersey devised a technique to look for organisms that had antibacterial properties. He prepared soil samples with their live bacteria and placed them in a Petri dish. He then placed an agar suspension on top of them containing pathogenic Gram-negative bacteria. He found that one soil bacterium in particular, the Gram-positive *Streptomyces lavendulae*, was very effective at inhibiting the growth of the pathogenic bacteria. The antibacterial chemical it produced, streptothricin, when purified, was found to be soluble in water and stable. It was also active against many Gram-negative bacteria as well as Gram-positive bacteria.

By discovering the genus *Streptomyces*, Waksman had identified the source of what was going to produce two-thirds of all our antibiotics.

There were toxicity problems with streptothricin which rendered it unsuitable for human use, so Waksman and colleagues examined a large number of *Streptomyces* species. Many produced antibacterials and the yield could be improved by altering the nutrients on which they grew. Most of these antibacterials, however, were toxic. In 1943, they found two strains that produced antibacterials in much larger quantities than they had seen before. The antibacterial was highly effective against most Gram-positive and Gram-negative bacteria and was capable of curing systemic infections in animals, without apparent toxic effects. They named the antibacterial streptomycin.

Streptomycin was as important an antibiotic discovery as penicillin. Firstly, Gram-negative bacteria could now be treated and these comprised a large proportion of the bacteria responsible for hospital-acquired infections. Waksman's colleague, Albert Schatz, tested streptomycin against *Mycobacterium tuberculosis* and found that it was inhibited. Streptomycin was not too toxic in animals infected with tuberculosis and this antibiotic eradicated the infection, even if they had been close to death prior to treatment. Pulmonary tuberculosis was first cured in a woman treated empirically in the USA over a six-month period in 1944.

The incongruity of these three major discoveries, penicillin, sulphonamides, and streptomycin, was that they were selectively toxic, in other words they preferentially attacked the bacteria without harming the human cells, but the reason was unknown. The initial discovery of penicillin had been serendipitous and luck had also played a considerable part in the discovery of sulphonamides and streptomycin. These drugs were much safer than those tried before but no explanation for this could be found. Bacteria from the *Streptomyces* genus were isolated all over the world and literally hundreds of antibacterial compounds were found; however, most were toxic to humans and could never have a role in therapy. There were only a few that were not, including chloramphenicol, tetracycline, and vancomycin—all antibiotics that we still use today. Other antibacterials soon followed: amphotericin, cycloserine, erythromycin, gentamicin, kanamycin, lincomycin, neomycin, nystatin, rifampicin, and spectinomycin; many of these also from the bacteria of the *Streptomyces* genus.

Fungi had produced penicillin, so other strains were also tested but they were not nearly as prolific as *Streptomyces*. Giuseppe Brotzu noticed in 1945 that the incidence of typhoid fever, prevalent in Italy at the time, was lower in his city of Cagliari in Sardinia than elsewhere. There was a raw sewage outlet but still those bathing near it did not get infected. Brotsu isolated a fungus, then called *Cephalosporium acremonium*, which clearly produced a substance that killed Gram-negative bacteria. Unable to analyse it further, he sent the strains to the group at the University of Oxford, who found they produced a myriad of antibiotics they called cephalosporins. One in particular, cephalosporin C, was suitable as a therapeutic agent.

Bacteria and fungi were not providing antibiotics for our benefit but rather for their own survival. These antibiotics are often called 'secondary metabolites'. They are not required for the growth of the organism but rather to promote its survival in certain locations. The fact that the source of these antibiotics was primarily soil-dwelling bacteria was no accident. The soil is a rich environment of competing bacteria and fungi, with a finite supply of nutrients. In order to thrive, many microorganisms have to inhibit the growth of the opposing bacteria and fungi. Many evolved metabolic pathways that produce compounds toxic to competing bacteria and then export them into their surroundings to ensure that their 'territory' is free of challengers. Thus it is not surprising that the vast majority of these compounds are toxic not just to other microorganisms but also to us.

By the start of the 1950s, most of the antibiotics derived from these soil organisms had been discovered. There were still bacteria that were not easily treated by these antibiotics and some antibiotics, including penicillin, could only be given by injection because they could not survive the hydrochloric acid in the stomach. In production, the yield had been increased by selecting certain strains of antibiotic-producing microorganisms and by altering the conditions of fermentation, particularly changing the nutrient source. Selective chemical modification of the antibiotics produced by fermentation could improve not only the penetration of the drug but also the spectrum. The best example is the modification of Fleming's original penicillin molecule, where the addition of an amine group not only prevented the molecule being destroyed by the stomach acid, so that it could now be administered orally, but also increased the spectrum so that it could now attack Gram-negative bacteria as well as Gram-positive. The antibiotic created was ampicillin, which became the most widely used antibiotic in the world (Figure 16).

The group of antibiotics that have undergone the most chemical modification are the cephalosporins, more than fifty different chemical modifications having been made, starting with the original molecules identified in Oxford. This has made them the most versatile group of antibiotics, which can be 'tailored' to deal with almost all bacterial pathogens.

Chemical modification still relied on fermentation to produce a natural product. Surely it would be easier to devise a chemical from basic principles and engineer it to inhibit specific pathogens. This has proved to be an elusive goal. In the late 1940s, George Hitchings of Burroughs-Wellcome was working on chemicals that he thought could inhibit the bacterial enzyme dihydrofolate reductase. He selectively modified parts of his core molecule and found one, pyrimethamine, that was effective against malaria. Encouraged by this success, he made further modifications and found another that was selective particularly against Gram-negative

Penicillin G
Obtained by fermentation

6-Aminopenicillanic acid
Obtained by chemical or
enzymic removal of most
of side chain at 6 position

Ampicillin
Obtained usually by
chemical addition of
the active side chain
at the 6 position

16. Diagram showing the stages of the chemical modification of penicillin

bacteria. This he called trimethoprim, which was put into clinical use in the 1960s. This was possibly the only case where a scientist actually engineered a completely new compound to inhibit bacteria—for which, like Fleming, Chain, Florey, Domagk, and Waksman before him, he earned a Nobel Prize.

The other major group of chemical antibacterials has been the quinolones and naphthyridines. The first, nalidixic acid, was discovered in the early 1960s by George Lesher from his work on chloroquine production. A relatively ineffective antibacterial, the activity of this group of compounds could be radically improved if a fluorine atom was placed at position 4 of one of the aromatic rings. Many derivatives were synthesized and marketed, though the most successful was ciprofloxacin. The source of natural antibiotics ran out more than fifty years ago and we have since relied on chemical modifications of these natural products or the few chemical antibacterials, such as trimethoprim and ciprofloxacin, ever since. However, the options for introducing new modifications to current compounds are virtually exhausted.

Pharmaceutical companies have, until recently, tried to devise new methods to discover antibiotics. One was to use high-throughput screening where they would examine almost all the chemicals ever produced to see whether any had any useful antibacterial activity. This was not a directed approach and was largely unsuccessful. The advances in molecular biology meant that the DNA of complete bacterial genomes could be compared with the information from the human genome project to identify targets in bacteria that did not exist in mammalian cells, thus allowing the possibility of devising a compound that would selectively inhibit the bacterium. A computer would screen the DNA sequences and highlight likely candidates. This approach also failed, and while it has been estimated that there are probably just over a hundred possible targets in bacteria, which do not exist in human cells, none of them could successfully be inhibited by chemicals. A further approach was to use the computer to model molecules using 3D-spatial information on the likely target (usually an enzyme) so that virtual chemical modifications could be added to a compound to improve its binding. This approach has also failed, which has meant that we now face the immediate future with the antibiotics that we currently have, or very close analogues.

The difficulty in designing a new antibiotic is essentially twofold. The compelling reason to discover new antibiotics is because bacteria have become resistant to the ones we have and some of them are very resilient. Therefore, more powerful chemicals are needed to control them. The conundrum is that more powerful antibiotics are more likely to be active against human cells as well as bacteria. This becomes more acute as national regulatory authorities are continuously upgrading the safety requirements of their pharmaceuticals. It is estimated that many of the original antibiotics, if they were presented now as new pharmaceuticals, would not pass the safety regulations. The fluoroquinolone trovafloxacin was removed from clinical use in 1999 because it had been associated with the deaths of fourteen patients who had suffered liver failure. At that time, about 300,000 prescriptions for trovafloxacin were being given each month in the USA. About 10 per cent of the population has an allergy to penicillin and should avoid its use; however, in the USA, up to 400 people a year die from anaphylaxis following penicillin use. There is no suggestion that penicillin should be removed from clinical use. A significant barrier to discovery and eventual clinical launch of new antibiotics is the safety requirements to which new antibiotics have to adhere. This is understandable when considering the use of antibiotics for simple community-acquired infections but is a little surprising when the treatment of life-threatening infections, particularly in hospitals, is considered and few viable antibiotics remain.

A patient who undergoes chemotherapy to try to remove a malignant tumour is given a drug that is essentially toxic to all human cells but is hopefully more toxic to the tumour cells. Some cancer chemotherapeutic drugs were so toxic that the patient had subsequently to be rescued with antidote drugs. Similarly, the treatment of HIV infection involved the use of potentially toxic drugs that inhibit enzymes in human cells. One, zidovudine, was originally considered as a candidate antibiotic. There is no question that these toxic drugs should not be given—because the severity of the patient's condition means that the potential benefit outweighs the risk. The same risk–benefit analysis should also be applied to the treatment of severe bacterial infections; there are many cases which will prove fatal if not treated but might be resolved with antibiotics that do not pass current safety regulations.

Selective toxicity

This leads to the question of how antibiotics work in the first place. Their essential characteristic is that they should be selective. The degree of selectivity has initially been quantified in a rather gruesome analysis of measuring the dose required to prevent the infection developing and the dose required to kill the animal, usually a group of mice that have been infected. This is called the *Therapeutic index* or *ratio*; it has been modified in humans as the ratio of the concentration of drug leading to toxicity of the drug divided by the minimum curative dose. As the ratio rises, the antibiotic is more selective. This measure of selective toxicity can identify how much antibiotic can be given and the likely success. The most toxic have ratios of eight whereas the least toxic have ratios of a hundred or more.

Although the actions of the two earliest antibacterial drugs, sulphonamides and penicillin, were not known at the time of their clinical launch, they were subsequently found to be selective because they inhibited enzymes that only exist in bacteria, dihydropteroate synthetase and transpeptidase respectively. Therefore, in theory, unlimited amounts of these drugs could be given as they should have no target to bind to in human cells. This is not the case; both drugs have been associated with severe adverse reactions: anaphylaxis for penicillin and Stevens–Johnson syndrome for sulphonamides. However, the ability to target a metabolic pathway that only exists in bacteria is an attractive model for antibiotic use. Penicillin's and cephalosporin's inhibition of transpetidase is at the final stage of the synthesis of the bacterial cell wall, a structure composed of subunits that are almost unique to bacteria, including sugars such as N-acetyl-muramic acid and the D-amino acids, for only L-amino acids are found in most living cells. Other antibiotics were found to inhibit the pathways leading to the synthesis of the cell wall, notably vancomycin, the basis of front-line treatment of MRSA. Binding primarily to the D-amino acids of the cell wall, vancomycin is only partially selective as it is quite often associated with adverse effects.

The main metabolic area where antibiotics have proved effective is the inhibition of the manufacture of proteins. Although the basics of this protein synthesis are the same in human and bacterial cells, the biological machinery is different. In both, protein synthesis occurs on the ribosomes, twin subunit molecules comprising both proteins and RNA, but bacteria and human cells have different compositions of these proteins and some antibiotics exploit this. These antibiotics, such as streptomycin and chloramphenicol, bind only to the bacterial ribosomes and not to those in human cells, thus selectively inhibiting bacterial growth. Some, such as tetracycline, can inhibit both bacterial and mammalian protein synthesis but are actively transported into bacterial cells and are thus excluded from mammalian cells; in this way they are selective. If the concentration of any protein-synthesis-inhibiting antibiotic is sufficiently high, they will start to prevent protein synthesis in the mammalian cells. So the Therapeutic index is important for this type of antibiotic. There are also antibiotics that inhibit bacterial folic acid synthesis, DNA and RNA synthesis, preferentially inhibiting the bacterial enzymes.

Although the targets of the main antibiotics are known, it is not always clear how they actually work. All bacterial cells have a rigid cell wall and this controls the osmotic pressure within the cell, which is generally high. Most of the cell wall synthesis inhibitors, but not all, leave the bacterial cell with an inadequate barrier to protect it against its environment and thus their action leads to the death of the cells. Antibiotics such as these are called bactericidal antibiotics. The DNA and folic acid synthesis inhibitors either directly or indirectly lead to incorrect synthesis of new DNA, causing the cell to die, and so they are also bactericidal. On the other hand, the inhibitors of protein synthesis, except for drugs like streptomycin, for the most part merely stop the bacterium from growing; these are known as bacteriostatic antibiotics. The relative clinical importance of each will be discussed below.

It could be argued that the better of these two types of antibiotic *must* be those that kill the bacteria. However, it was found early on that infections responded equally well to both bactericidal and bacteriostatic antibiotics. The rationale behind this is that the main role of an antibiotic is to prevent further multiplication of the bacteria causing the infection, thus limiting the bacterial load, which would surely increase without the administration of the drug. Once the progress of bacteria has been halted, the immune system could take over and deal with what had effectively become a static system. This model does work well for infections in patients with an efficient immune system and early antibiotic therapy was based on the assumption that the patient would finally eradicate the bacteria. Since the 1960s, changes in medical and surgical practice have required greater capability from antibiotics. Transplantation of organs requires the administration of immunosuppressive drugs to combat rejection. Similarly the use of certain cancer chemotherapeutic agents reduces the white blood cell count, known as neutropenia, which also lowers the ability to combat bacterial infection. Both types of patient are increasingly vulnerable to bacterial infection as they have very limited capacity to eradicate the bacteria. Of course, they are administered antibiotics, often aggressively, to limit the onset of infection; however, when they are infected the role of the antibiotic merely to stop the bacterium dividing can become inadequate as there is no capacity to 'mop up' remaining cells, and infections can persist. Thus, for these types of patient, an antibiotic that can kill the bacteria would be preferable to one that merely stops the bacteria from dividing. Indeed, almost all the antibiotics used in this situation are bactericidal and can kill bacteria quickly. Another advantage is that a bacteriostatic antibiotic would allow large numbers of bacteria to remain and might promote the emergence of resistance; killing the bacteria reduces this risk as 'dead bacteria cannot become resistant to antibiotics'.

There can be disadvantages to killing the bacterium, especially if the cell wall is broken in the process. There may be toxins either within the cell that are released or which form part of the cell wall (endotoxins), which now become soluble within the bloodstream, sometimes leading to septic shock. This is particularly apparent with Gram-negative bacteria.

The ability of an antibiotic to lyse the cell (colistin) or produce an inadequate cell wall (penicillin) is an obvious cause of cell death; however, these are the exceptions. There are four main groups of bactericidal antibiotics: the β-lactams (such as penicillins and cephalosporins), the fluoroquinolones (such as ciprofloxacin), the diaminopyrimidimes (such as trimethoprim), and aminoglycosides (such as streptomycin). Some β-lactams do not lyse the cell wall but rather prevent the cell division; in this case they behave much more like the other groups. Each group inhibits a different step of bacterial metabolism but they all kill bacteria in approximately the same way. At the time that the antibiotic acts on the bacteria, protein synthesis must take place. If it is prevented, then the antibiotics do not kill. When the proteins that are produced by the action of each group of antibiotics are analysed, they are often very similar: a combination of 'shock' proteins that were originally identified when bacteria were exposed to extremes of temperature. These proteins encourage a survival response which causes the death of the majority of bacterial cells, releasing vital metabolites that the surviving cells can use to rescue them. A small proportion of cells in most bacterial populations are metabolically inert at any one time and thus do not produce these proteins; these are likely to be the potential survivors. In other words, the antibiotics elicit a response encouraging the majority of bacteria to kill themselves. If the concentration of antibiotic is sufficiently high, then there may be no survivors, but sometimes survivors persist and they are able to proliferate again once the antibiotic has been removed.

Quorum sensing

It has been known for some time that bacteria within a colony of cells are able to communicate with each other; this is known as quorum sensing. The ability to do this depends on the number and concentration of bacteria present. Some bacterial cells within the colony produce signalling molecules. These are N-acyl homoserine lactones in Gram-negative bacteria and often oligopeptides in Gram-positive bacteria. Other bacterial cells in the colony have receptors which detect these molecules. The binding of these signalling molecules at the receptor induces the activation of certain genes. The larger the number of bacterial cells, the more signalling molecules are able to bind to the receptors and the greater gene activation. If the cell concentration is too low, the binding of the signalling molecules is insufficient to trigger the gene activation (Figure 17). Furthermore, once the concentration of signalling molecules reaches a threshold it induces even more of its own production. So many bacteria have the ability not only to communicate but, with an indication of their numbers, they can initiate different collective responses (Figure 17). These responses can control many collective bacterial functions, such as cell division. We are not yet fully aware of all

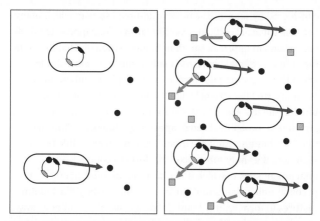

17. **Quorum sensing. Left, bacterial density is low so there are few signalling molecules or autoinducers (black dots) failing to reach a threshold. Right, bacterial density is high with many autoinducers, which switch on genes for specific substances (grey squares) needed when large numbers of bacteria are present**

the bacterial functions affected by quorum sensing but the response to antibiotics may be one of them. As stated above, not all bacteria in a colony are equally destroyed during antibiotic challenge and the lack of protein synthesis may be the factor. If quorum sensing can control the stages of cell division it will also be able to control whether certain proteins may be manufactured and so the population of bacteria may control its response to the antibiotic. There is much evidence to show that the success of an antibiotic is dependent on the concentration of bacteria challenged: the more bacteria present, the less effective the antibiotic.

Biofilms

The requirement for protein synthesis also saves bacteria in another situation. Many bacteria, including some pathogens important in human infection, can form biofilms. They are considered by some as the most widespread state in which many bacteria can exist. Although biofilm formation happens regularly on surfaces in nature, a common manifestation of a biofilm is the formation of dental plaque on teeth. Biofilms become serious when they occur on inert devices, such as plastic tubes, that are often inserted in extremely ill patients. In the formation of a biofilm, the first bacteria that encounter the inert aqueous surface attach to it, probably through appendages known as fimbriae or pili. This binding strengthens and becomes irreversible by the secretion of a slimy substance. This forms a complex mix of attachment sites to which other bacteria can bind. The cells are probably able to communicate by quorum sensing. The biofilm matures to form layers of non-dividing cells, often but not always of a single species. The biofilm is covered by more actively dividing (planktonic) cells (Figure 18). The formation of biofilm has advantages in that the bacteria, in close proximity, have a heightened ability to communicate and they can release planktonic cells to colonize other environments. It is, however, their response to antibiotics that is of most concern. Bacterial cells in biofilms have a vastly increased capability to overcome the effects of antibiotics, perhaps decreasing their susceptibility 1,000-fold. Two factors work against the antibiotic: the first is that biofilms are a mass of bacteria through which the antibiotic has difficulty in penetrating; the second is that the inert environment and lack of appropriate protein synthesis ensure that the bacteria survive for longer.

Spectrum of activity

Some early antibiotics had a narrow spectrum of activity but with the advent of the broad-spectrum antibiotic streptomycin, almost all bacteria could be controlled. This raises the issue of which antibiotic is better, broad or narrow spectrum. It is an extremely difficult question to answer. A broad-spectrum antibiotic will almost certainly start inhibiting not just the pathogenic bacteria but most of the non-pathogenic bacteria. This can have mild

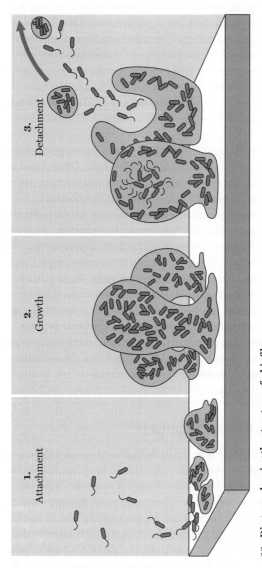

18. Diagram showing the structure of a biofilm

adverse effects, such as diarrhoea, or the more serious eradication of much of the gut flora and super-infection, with *Clostridium difficile* causing antibiotic-associated pseudomembranous colitis. This can become life threatening and is the reason why administration of the broad-spectrum cephalosporins is now often restricted. Therefore administration of a broad-spectrum antibiotic can cause complications in areas of the body distant from the site of infection.

Conversely the administration of a narrow-spectrum antibiotic may be as detrimental for the opposite reason. There have been cases where narrow-spectrum antibiotics have been used to treat chest infections caused by the Gram-negative bacterium *Haemophilus influenzae*. Although excellent for the eradication of Gram-negative bacteria, they were without effect on Gram-positive bacteria. *Streptococcus pneumoniae*, a Gram-positive bacterium commonly causing pneumonia, can be an underlying infection kept in check, in part, by the less pathogenic *Haemophilus influenzae*. Eradication of the latter, however, can suddenly provide a bacterium-free environment in a vulnerable patient with active selection for the new pathogen. The essence of good prescribing is to provide an antibiotic with a sufficiently broad spectrum to deter any other bacteria entering the cleared site of infection without eradicating large numbers of bacteria in other parts of the body.

Oral or injectable antibiotics

The site of administration of the antibiotic has an influence on this. There is a preference in the United Kingdom and North America for the oral administration of antibiotics, whereas elsewhere in the world antibiotics are given, through choice, by injection. The latter does allow more rapid accumulation of the antibiotic at the site of infection, as it is quickly taken by the bloodstream. On the other hand, oral antibiotics have to pass through the stomach acid and become absorbed in the small intestine. This is a circuitous route and dependent on the efficiency of absorption, which is surprisingly poor for some antibiotics. This can result in much of the available antibiotic passing straight through the gastrointestinal tract, killing many of the resident harmless bacteria and providing an environment for the selection of resistant strains.

Why antibiotics may fail

Antibiotics fail to cure an infection because we are trying to treat bacteria in a biofilm or the causative organism has become resistant (next chapter). There are other fundamental reasons as well. The delivery of an antibiotic to the site of infection ultimately depends on the bloodstream, from which the antibiotic will diffuse out of the capillaries. Many infections, including wounds, are often surrounded with an almost impenetrable layer of pus and other discarded products of the immune system. These are difficult matrices through which the antibiotic must pass in sufficient concentration to eradicate the infection. In these cases, cleaning or debridement of the wound is at least as effective as antibiotic treatment and success is often not achievable without it. Often wounds are cleaned with antiseptics, which are less selective than antibiotics and can kill the patient's cells required for regeneration.

Chapter 7
Antibiotic resistance

Ever since the introduction of antibiotics, bacterial resistance has been an issue; however, it is not until recently that this issue has become one of major importance, attracting the attention of the World Health Organization, policymakers, and politicians. The essential reason for the change is that we are exhausting our arsenal of new antibiotics. Many pharmaceutical companies are now reluctant to invest in new antibiotics and the difficulties in discovering new antibacterial drugs that are safe enough to be given to clinical patients are immense. While there is no shortage of antibacterial compounds, the vast majority of them would fail current safety standards. The problem was perceived to be driven by the Gram-positive bacteria, such as MRSA, but actually the acute crisis is the continued capability to treat infections caused by Gram-negative bacteria.

Mutations in the bacterial chromosome

Bacterial infections are composed of, literally, millions of individual cells. These may be in a liquid environment such as the blood or urine or, perhaps, more likely they are congregated at a focus of infection. These bacteria are not really individual cells but, as we have seen, they form a colony of cells with, largely, an identical set of genes. The survival of the colony requires cooperation between the individual cells and the 'sacrifice' of the majority so that the 'gene pool' can continue to exist. The simplest form of resistance comes from the ability of mutations to occur within genes. This has an average frequency of approximately 1 in 10^7 bacterial divisions for any single gene. Thus within one billion bacteria, which make such a colony, there might be a hundred bacteria that have undergone spontaneous mutation in a specific gene and this mutation may confer resistance to a specific antibiotic. When this population is treated with the antibiotic, the vast majority of the bacteria, all but the hundred with the mutation, are inhibited or killed but the mutants are able to prosper and they start dividing. They can soon replace the original infection site with their progeny. Mutations occur during DNA replication prior to bacterial division, which means it is theoretically possible to have a population of bacteria with no resistant mutants when antibiotic treatment begins; however, if the concentration of antibiotic is insufficient, DNA replication will continue and mutants can form. This type of mutational resistance is the reason why doses of antibiotic have to exceed a concentration threshold to prevent the mutants surviving; this has more recently been described as the 'Mutant Prevention Concentration' (Figure 19).

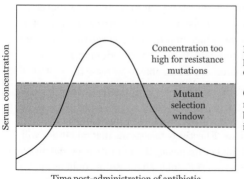

Time post-administration of antibiotic

19. Mutant Prevention Concentration

Once a single mutation has occurred and the resistant bacteria have taken over the infection site, it is possible for a second mutation to occur by the same process, giving an even higher degree of resistance, and the progeny of this double mutant will divide and then take over the population. The gene encoding the target for a given antibiotic is usually the gene in which the mutations take place; for instance, the *gyrA* gene encodes a subunit of the enzyme DNA gyrase to which the antibacterial drug ciprofloxacin binds. Sequential mutations in the *gyrA* gene produce a DNA gyrase molecule that binds ciprofloxacin far less tightly and thus confers resistance. The original *gyrA* gene encodes a protein with a structure that has evolved over an aeon and presumably this has resulted in the most efficient structure for the role it performs. While the mutations still allow the original role to occur, it may well do so less efficiently than the non-mutated protein. This means that there is an efficiency 'cost' for the bacterium to carry the mutation and often these bacteria divide more slowly in the absence of antibiotics than the bacteria without the mutant. This results in pressure, once the antibiotic has been removed, for reverse or back mutations to occur and this is often seen. This is a reason why strategies to overcome bacterial resistance have suggested that the restriction or banning of certain antibiotics could reverse the development of resistance; however, back mutations are not the only mutations that can occur under these circumstances. Mutations can occur in other sites in the genome, which improve the division rate; these are called compensatory mutations. The bacterium retains its resistance mutations and is also able to decrease the 'cost' that these mutations have caused.

Some strains of bacteria are able to pass into a transient hypermutator state. Effectively this means that mutations will occur at a much higher frequency than 1 in 10^7 bacterial divisions, maybe an increase of 1,000-fold. Whereas this could be an advantage in a situation where the bacteria are likely to be challenged with antibiotics, this mutation rate will occur in all genes producing disadvantageous mutations at least as often as those giving an advantage. Disadvantageous mutations are almost always lethal, so the hypermutator state is often precarious for a bacterial population and they often pass back into their normal state. The capability of increased frequency of mutation is found as a permanent feature in some strains of bacteria that have become resistant to many antibiotics and is thought to have been a contributory factor to their multi-resistant character. Although mutation has been crucial to the development of all resistance mechanisms at some point, mutations emerging in the chromosomes of bacteria treated with antibiotics are a surprisingly rare form of resistance.

Plasmids and mobile structures

The Japanese were the first to notice that some bacteria became resistant not to just one antibiotic but to four at the same time. This could not be explained by chromosomal mutations as it would have required a mutation frequency of 1 in 10^{28} bacterial cell divisions, producing a mass of bacteria greater than that of the moon. Clearly this did not happen and the resistance genes were subsequently found to have been imported en bloc by a plasmid. These circular molecules of DNA are independent of the bacterial chromosome and have their own replication origin. The plasmids conferring resistance in clinical bacteria usually replicate at the same time as the bacterial chromosome, so that the bacterium has approximately two to five copies of the plasmid. They have the genetic machinery that allows them to transfer from one bacterium to another; this is actually an extremely difficult process and requires approximately twenty-five genes but it allows the plasmid to pass from one strain to another and from one species to another, even travelling to quite different bacterial species. The constraint on unlimited transfers between bacteria is the defence mechanism that bacteria have evolved to protect themselves from incoming DNA. They produce endonucleases that cut and degrade incoming DNA and the plasmid has to overcome these restriction enzymes in order to survive in the new bacterium. The closer the new host is to the old one, the less restriction will occur. These endonucleases or restriction enzymes are able to recognize specific short sequences of DNA and, as a by-product of this capability, they are now a crucial element in many molecular biological procedures.

Chromosomal resistance genes are usually under a repression system so that they are not activated, whereas those on plasmids are usually maximally expressed to give the highest levels of resistance continuously. Re-examination of some of the earliest bacteria ever identified shows that the plasmids themselves have been present for a long time, though the presence of resistance genes is, for the most part, a recent development. The genes conferring resistance are often clustered together on plasmids, indicating that there is a preferred location. This gives a clue as to how they were captured by the plasmid. Bacterial DNA is littered with elements known as insertion sequences. These are relatively short, about 700–2500 nucleotide base pairs long, mobile elements of DNA that can move from one area of DNA to another. At the extremities of the insertion sequence are often inverted repeats and between these there are often genes that encode enzymes facilitating movement of the element. Similar insertion sequences can often align themselves closely together on the same sequence of DNA. When this happens, the whole region of DNA, including the two insertion sequences and the region in between, can be moved. If there are other genes, in this case resistance genes, between the insertion sequences, this composite element is known as a transposon.

Often transposons insert into specific sites within DNA and this is commonly the case in plasmids. By the successive introduction of resistance genes, sited in transposons, into a particular location within the plasmid DNA, a cluster of resistance genes form. Often the structure of the original insertion sequences of the early transposons is disrupted by further transposition events, bringing in new resistance genes. This fixes the early resistance genes in the plasmids and prevents them being transposed 'out of' the plasmids. Insertion sequences can, in theory, surround and pick up any gene in a transposon; however, selection plays an important role and the reason why we find clusters of resistance genes is because antibiotic usage has selected the plasmids that have particular resistance gene combinations. Transposons along with plasmids provide a symbiotic relationship with the host bacteria. They often carry accessory genes, resistance in this case, that are beneficial but not usually essential for the survival of the organisms. If they are lost, the bacterial population could still survive. Often, however, the exposure to antibiotics is so great that it is more beneficial for the bacterium to carry the resistance gene in the chromosome than on a plasmid. Transposons carrying resistance genes are increasingly found located in the bacterial chromosome.

It could be argued that this is not the evolutionary end point for resistance genes because many multi-resistant bacteria that are spreading clonally through the clinical populations retain their resistance genes on identical plasmids. It is likely that there is a 'match' between certain plasmids and bacteria. Plasmids, as extra-chromosomal DNA, do provide an energy drain on the resources of the cell. In some successful cases, where a plasmid has established itself in a particular pathogen, the cost has been ameliorated by mutations in the plasmid and the host DNA, ensuring that this becomes a favoured combination. There is a further advantage for the plasmid as not every bacterial cell within the population needs to carry the plasmid, though they often do, because if the bacteria are suddenly challenged with antibiotics then only those with the plasmid survive.

The capture of resistance genes by insertion sequences is a relatively inefficient process and does not alone explain the rapid acquisition of resistance genes by transposons and plasmids. Within each are often found integrons. These are short DNA sequences that can capture genes by encoding an integrase enzyme. The gene, *int1*, encoding integrase, is closely linked to an attachment site (*att*), into which captured genes are inserted, and a promoter, from which the inserted gene is expressed. The gene cassettes that integrons capture often do not confer resistance, but antibiotic usage ensures that many of those integrons identified in clinical bacteria do carry resistance genes.

Mechanisms of resistance

Mutation of the target of an antibiotic is not an option for plasmid-mediated resistance. The resistance mechanism has to be dominant and is usually, though not always, the result of a single gene. Almost all bacteria possess genes encoding enzymes that export often unwanted metabolites (Figure 20). When encoded by the chromosome, these efflux pumps are usually under strict repressed control and are activated only when needed. Some of them are able to efflux antibiotics, thus providing a low level of resistance as they export the drug faster than it can enter. Some plasmids have acquired genes for efflux and are able to confer resistance to antibiotics such as tetracycline. The efficiency is dependent on the concentration of antibiotic outside the cell and, as more is being imported, the efflux pump is soon overwhelmed.

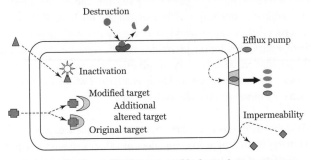

20. Major mechanisms of resistance used by bacteria to overcome antibiotics

Thus they give relatively low levels of resistance and are, in many cases, just the first line of defence. Clinical concentrations of antibiotics are high and the resistance mechanisms must match this. The most widespread mechanism is inactivation of the antibiotic and this accounts for the vast majority of resistance genes identified. This is manifested in two ways: by the enzymic destruction of the antibiotic or by the addition of a chemical group that is added to the antibiotic so that it is no longer taken up or recognized as an active antibiotic.

Enzymic destruction is exemplified by just one enzyme, the beta-lactamase. Many bacteria have genes encoding them on their bacteria chromosome but in clinical bacteria, the infiltration of beta-lactamases has been on plasmids. There are now more than 1,000 identified in clinical bacteria. The enzyme attacks the beta-lactam ring, which is crucial to the function of pencillins, cephalosporins, and carbapenems. Beta-lactamases are a prime example of convergent evolution, as four distinct molecular classes, A–D, have been identified. Class A is the chromosomal beta-lactamases of Gram-positive bacteria and has evolved primarily to confer resistance to penicillins. Class C is the chromosomal beta-lactamases from Gram-negative bacteria and its primary target are the cephalosporins. The class D beta-lactamases are chromosomal beta-lactamases of one genus of Gram-negative bacteria, *Acinetobacter*. All three classes have the same basic mechanism of action, using a serine residue for their enzyme catalysis, though their overall structures are completely different. Class B beta-lactamases have a different mechanism of action; they are metallo-enzymes, so use a zinc ion to catalyse the destruction of the antibiotic and they are particularly effective against the carbapenems. The origins of this class of beta-lactamase are unknown.

The class A beta-lactamases were transported from their Gram-positive origins to the plasmids of Gram-negative bacteria. The most prevalent is called TEM-1 after the little girl, Temoira, who provided the bacteria from which it was first identified. This beta-lactamase was unknown before the introduction of ampicillin and has now become the most widespread antibiotic resistance mechanism in the world, so much so that a quarter of the population carry commensal bacteria containing this resistance gene. The TEM-1 beta-lactamase provided a very efficient resistance to the pencillins but was largely unaffected by the cephalosporins; therefore there was a massive rise in the number of new cephalosporins capable of overcoming this resistance. However, in 1982 there was an outbreak in a neonatal unit in Liverpool, England. The causative bacterium, *Klebsiella oxytoca*, was treated with the cephalosporin ceftazidime until the bacterium became resistant. The TEM-1 beta-lactamase that it was carrying had mutated at position 164 with the effect of opening the active site and allowing ceftazidime to be destroyed. Once this first mutation had taken place other mutations soon followed, until well over a hundred different enzymes all derived from the original TEM molecules have been found and virtually all cephalosporins could now be resisted. These were known as extended-spectrum beta-lactamases, or ESBLs. A Class A beta-lactamase similar to TEM though far less common is SHV-1; it has the same activity against pencillins. The SHV-1 beta-lactamase also started to mutate until it also produced over a hundred ESBLs. This had the effect of not only reducing the usage of cephalosporins but also restricting their further development as new antibiotics.

Actually the proportion of ESBLs based on TEM and SHV are now declining, as a new group of Class A ESBLs is emerging. These are the CTX-M ESBLs, CTX standing for cefotaxime, the cephalosporin antibiotic that is the primary target of these enzymes. Unlike the plasmid origin of TEM and SHV, the CTX-M enzymes derive from the chromosomal beta-lactamases of *Kluyvera* species, which are closely related to *Escherichia coli*. The CTX-M-3 beta-lactamase is identical in structure to the chromosomal beta-lactamase in *Kluyvera ascobata* and appears to have been transferred directly from that species via a transposon and plasmid. CTX-M-3 is relatively rare in clinical bacteria but a single mutation at position 240 increases its capability to confer resistance to ceftazidime and is called CTX-M-15. CTX-15 is the most widely distributed ESBL in clinical bacteria in the United Kingdom and probably throughout the world. The transfer of CTX-M genes from *Kluyvera* appears to have happened on five occasions and has provided five distinct genetic subgroups of CTX-M beta-lactamases, each group named after the first enzyme discovered. The CTX-M-1 subgroup, which includes CTX-M-3 and CTX-M-15, was derived from *Kluyvera ascorbata*; the CTX-M-2 subgroup from a different strain of *Kluyvera ascorbata*. The CTX-M-8 and CTX-M-9 subgroups were derived from two distinct strains of *Kluyvera georgiana*. The species that is progenitor of the last subgroup, CTX-M-25, is not known. The subgroups are not closely related to each other but the enzymes within each subgroup are very similar. Therefore, in each subgroup a single gene will have transferred into clinical bacteria from *Kluyvera* and then started to mutate; the selective pressure of different cephalosporin antibiotics in clinical use will have allowed a myriad of different CTX enzymes to emerge within each subgroup. After CTX-M-15, the most important of these is CTX-M-14, a member of the CTX-M-9 subgroup, as this has been the most prevalent ESBL in clinical bacteria from the Iberian peninsula and, in the United Kingdom, in bacteria isolated from farm animals. This probably reflects the antibiotic usage in these areas.

The decline of the cephalosporins has produced a greater clinical reliance on the carbapenems. These antibiotics are, by and large, immune to the ESBLs, although there is a notable exception for strains containing the CTX-M-15 beta-lactamase, which can become carbapenem resistant as long as other resistance mechanisms, such as efflux pumps, are working in concert. Carbapenem resistance has come from new Class A beta-lactamases such as KPC-1 in *Klebsiella pneumoniae*. However, the largest group of plasmid-encoded beta-lactamases able to confer resistance to carbapenems come from Class B. Two large groups of closely related beta-lactamases, IMP and VIM, have evolved over the last decade or so to confer carbapenem resistance. They are efficient but the greatest concern has been expressed about the emergence of another Class B beta-lactamase that was originally identified in New Delhi, India, called NDM-1. The gene encoding this beta-lactamase has the ability to migrate into many different bacterial species and can confer high-level carbapenem resistance.

The prototypes of the beta-lactamases found in clinical bacteria did not evolve entirely during the time that we started using antibiotics. They were initially defence mechanisms used by soil bacteria to gain an advantage against other bacteria and fungi such as *Penicillium notatum*. Without these enzymes, these bacteria would never have been able to survive. They have then been transported to clinical bacteria by successive, and hitherto unknown, interactions between environmental and clinical bacteria; in the latter, mutations have been selected by successive challenges with different antibiotics.

Many resistance mechanisms originally derive from the bacteria that originally produced the antibiotic. The antibiotic is used by the bacterium to destroy other bacteria in the immediate environment but it runs the risk of killing itself. Therefore, it had to evolve a defence system. An effective method is modification of the enzyme that produces the final antibiotic. If the gene is duplicated and then mutates, then the resultant enzyme may bind the antibiotic and attach a functional group to it to render it ineffective (Figure 20). This is the basis of aminoglycoside modification. There are three enzymic groups: N-Acetyltransferases (AAC) catalysing acetyl CoA-dependent acetylation of an amino group; O-Adenyltransferases (ANT) catalysing ATP-dependent adenylation of a hydroxyl group; and O-Phosphotransferases (APH) catalysing ATP-dependent phosphorylation of a hydroxyl group effectively adding an acetyl, adenyl, or phosphate group respectively to various positions on the molecule, providing high-level resistance. There are over fifty of these enzymes mainly encoded on plasmids. More recently, a new enzyme has emerged, AAC(6')-ib-cr, which has the ability to add both an acetyl group *and* efflux fluoroquinolones—a very versatile enzyme!

Destruction and modification are the normal resistance mechanisms for antibiotics derived from natural sources; the bacteria face a greater challenge when the drug has no counterpart in nature. Trimethoprim and sulphonamides are completely synthetic and bacteria would never have previously been exposed to them, though resistance emerged within just a couple of years after the introduction of each compound. The initial resistance was caused by mutations in the target of the antibiotic, though this often only gave a low level of resistance. Bacteria that were highly resistant emerged soon afterwards and the genes responsible were found to have been introduced on plasmids. The mechanisms were different from those previously identified. The plasmids encoded an additional target enzyme: a dihydrofolate reductase in the case of trimethoprim resistance and a dihydropteroate synthetase for sulphonamide resistance. The additional target enzymes could still perform the task of the chromosomal enzyme but they were far less susceptible to inhibition by the antibiotics (Figure 20). Therefore they 'bypass' the antibiotic's inhibition of the chromosomal enzyme. Although this can produce very high levels of resistance, an increase of up to 10,000-fold, it is a fairly inefficient method of resistance as the bacterium has to expend energy to produce two enzymes, one of which is effectively useless as it is being inhibited. This mechanism can only be employed when there are relatively few original enzyme molecules produced by the cell, up to just ten in these two cases. The origin of these enzymes is not clear. Some came from mutations in chromosomal enzymes of closely related species, though these tended to give low levels of resistance. However, this process of sequential mutation is not sufficient to explain the rapid emergence of high-level resistance; the genes for these must have been present before the drugs were introduced. In the case of trimethoprim resistance, some resistant dihydrofolate reductases came from unknown sources that are unlikely to have been bacterial in origin. We know that almost all mammalian dihydrofolate reductases are resistant to trimethoprim, hence the selective toxicity of the drug, and it is conceivable that some of these mammalian genes may have been taken up by bacteria.

Since the discovery of these bypass mechanisms in 1974, it was thought that trimethoprim and sulphonamide resistances were the only examples. In the 1990s, there was a rise in vancomycin-resistant *Enterococcus* (VRE) strains that were becoming resistant to the final drug of choice, vancomycin. Prior to this resistance, the most common pathogenic species was *Enterococcus faecalis* but a greater proportion of vancomycin-resistant *Enterococcus faecium* emerged. The resistance was often high level and ensured the ineffectiveness of the antibiotic. Vancomycin is a natural product derived from the genus *Streptomyces* and it might be expected that the resistance mechanism would be destruction of the antibiotic. However, the mechanism is much more complex and is, by far, the most sophisticated antibiotic-resistant mechanism found so far. The binding site of the antibiotic is the D-alanyl-D-alanine dipeptide at the end of the cross-linking pentapeptide of the peptidoglycan polymer of the cell wall. During the cross-linking process, the bond between the two alanine residues of D-alanyl-D-alanine is broken, the terminal D-alanine is removed, and the cross-link occurs with the remaining D-alanine residue. The resistance mechanism exploits this excision.

Normal synthesis of peptidoglycan requires the ligation (binding) of two D-alanine molecules to form D-alanyl-D-alanine. The resistance mechanism comprises three main genes; the first reverses this reaction so that any D-alanyl-D-alanine formed by the cell will be converted back to two molecules of D-alanine. A second enzyme converts the product of the glucose metabolism, pyruvate, into D-lactate, and a third enzyme ligates the D-lactate to D-alanine to produce D-alanyl-D-lactate. The latter is then used instead of D-alanyl-D-alanine in the formation of the pentapeptide in the synthesis of peptidoglycan. Vancomycin does not readily bind to D-alanyl-D-lactate and so the cross-linking of peptidoglycan is not inhibited. When the cross-linking occurs, the bond between the D-alanine and the D-lactate is broken, the D-lactate is removed, and the cross-linking uses the D-alanine as before. So, the composition of the final molecule is exactly the same as before; only the means of production is different.

There are a number of variations of this resistance mechanism and the genes responsible can be carried on a plasmid along with control and non-essential accessory genes. Most other plasmid-encoded resistance mechanisms are a single gene product that is dominant within the cell; vancomycin resistance is different and takes over the cell's normal synthetic machinery. Such a sophisticated mechanism of resistance could not have evolved and have been refined in the sixty years that vancomycin has been used clinically. Actually very similar mechanisms have been found in *Streptomyces toyacaensis* and *Amycolatopsis orientalis*. Despite its name vanco-, meaning 'of unknown origin', vancomycin was originally discovered in a strain of *Amycolatopsis orientalis*, an actinomycete isolated from a soil sample in Borneo. The presence of three genes, in the same orientation as those found in plasmids in clinical *Enterococcus faecium*, suggests that they evolved in the species producing vancomycin in order to protect it from the antibiotic it was manufacturing. These genes were possibly captured by *Streptomyces toyacaensis*, another soil inhabitant, in order to protect this species against the drug. *Streptomyces toyacaensis* produces its own compound to attack neighbouring microorganisms but this is also a neurotoxin and too dangerous for clinical use.

Antiseptics and biocides

Most attention is directed at antibiotic resistance but actually we use a raft of other antibacterial chemicals, which we classify as biocides and subdivide as disinfectants and antiseptics. The difference between the latter two is often the concentration used: disinfectants are used at high concentrations, which can be toxic to Man, whereas antiseptics are used at much weaker concentrations as they are often applied to the extremities of the body. Should we be concerned about resistance to these compounds? The concentrations of disinfectants used clinically are often more than 1,000 times the concentration required to inhibit or kill the bacteria and they are usually indiscriminate, so will kill all microorganisms. Although resistance genes to these compounds do exist, resistance is unlikely to be a problem if the compounds are used appropriately. Even when resistance genes are present, the level of resistance they confer is far lower than concentrations that the bacteria are likely to encounter. Where problems are likely to exist is if the disinfectant is not used correctly (wrong concentration, inappropriate application, etc.). If a disinfectant is not removed completely after use and residues remain, bacteria that possess resistance genes may be able to proliferate. Similarly, if there is a large amount of organic matter, such as blood, then the efficacy of the disinfectant can be considerably weakened and resistant bacteria may proliferate; however, if the same bacteria are then placed in a solution of appropriately prepared disinfectant, then these 'resistant' bacteria should not survive.

The case is less clear with the antiseptics and those biocides that we incorporate into everyday objects such as cutting boards. Weaker biocides tend to be more discriminatory not just against bacteria but also the type of bacteria that they are able to inhibit. Generally speaking, they are better at controlling Gram-positive bacteria than Gram-negative. The antiseptics have more difficulty disrupting the double membrane and thicker cell wall of

Gram-negative bacteria than they do with Gram-positive bacteria. The concentrations used are often, but not always, higher than those required to kill the bacteria. Resistance genes have emerged and these generally confer resistance through an efflux pump. One widely used group of antiseptics are the quaternary ammonium compounds, which are particularly effective against Gram-positive bacteria. Resistance genes, called *qac* encoding efflux pumps, have emerged, sometimes in multiple copies. *qac* genes are often found in MRSA closely linked to the genes that confer antibiotic resistance, which leads to the speculation as to whether antiseptic use can select for antibiotic resistance. This has been further compounded with the discovery that *Acinetobacter baumannii*, an emerging hospital pathogen, can have up to four copies of the gene. In this case they were found closely linked to genes conferring heavy-metal resistance.

The presence of the *qac* and other resistance genes is not always associated with decreased susceptibility to antiseptics. This may, in part, be due to the fact that we do not have clear procedures for measuring antiseptic susceptibility such as we do for antibiotics. Furthermore, antiseptic preparations often contain mixtures of compounds, including surface active agents to promote the effect of the main component. However, it may simply be due to the fact that antiseptic resistance genes are not really what we suppose them to be and are actually selected by exposure to other compounds. The close association of *qac* genes with heavy-metal resistance genes in *Acinetobacter baumannii* suggests that these genes may have been selected when the bacterium was in the environment before it became a clinical pathogen.

Clinical resistance

Microbiologists measure bacterial susceptibility by observing the ability of the antibiotic to inhibit the bacteria, when growing on agar in a Petri dish. The simplest method is to spread a bacterial culture, which has been isolated from an infection, on the surface of agar in a Petri dish. The agar contains nutrients that allow the bacteria to grow. On the bacterial 'lawn' are placed small circular discs made of filter paper and containing fixed and accurate amounts of antibiotics. The antibiotic diffuses out of the disc into the agar. After the Petri dish has been incubated, usually at 37°C (normal human body temperature), the bacteria will have grown over the surface of the agar except around the discs of antibiotics, where there will be a circular zone of no growth where the antibiotic remains in sufficient concentration to inhibit the bacteria. The more susceptible the bacteria are to the antibiotic, the larger the zone. The size of the zone is usually correlated with the sizes of zones for the same antibiotic against control susceptible bacteria. The more resistant the bacteria, the smaller the zone so that very resistant bacteria may have no zone at all.

Disc susceptibility is a fast and reasonably accurate determination of bacterial susceptibility but it is not easy to quantify. In order to achieve this, the minimum inhibitory concentration of the antibiotic has to be determined. This is usually accomplished by placing the bacteria onto agar in Petri dishes containing increasing concentrations, usually in doubling increments, of antibiotic. After incubation, the lowest concentration that is able to prevent visible growth on the agar is known as the minimum inhibitory concentration, more commonly referred to by its acronym MIC. This is a very accurate determination of bacterial susceptibility. More recently modifications of these techniques have been made and there is much research into the ability to predict susceptibility and resistance by the genes that the bacterium carries. These do not involve culture of the bacterium but rather amplification of its DNA by the polymerase chain reaction (PCR). It is possible to identify not only specific genes but also whether they are being expressed. Depending on the combination of expressed genes in a particular species of bacteria it may be possible to 'predict' what the resistances would be. The advantages of these DNA-based techniques would be the speed at which they could deliver a result as there is no culture of the bacteria, and to some extent the cost, as fewer specialized personnel would be involved. The use of DNA-based techniques may seem fanciful, not least because they extrapolate in order to give a prediction; however, the culture-based techniques are also, in essence, just predictions.

So what do we mean by clinical resistance? Microbiologists are keen to report reduced susceptibility to an antibiotic or biocide but whether this can be translated in the failure of an antibiotic to cure an infection is a different matter. According to the old National Committee for Clinical Laboratory Standards (NCCLS) definition, 'The implication of the "susceptible" category implies that an infection due to the strain may be appropriately treated with the dosage of the antimicrobial agent recommended for the type of infection and infecting species'; thus the converse is true, that a bacterium is not clinically resistant if the organism can still be controlled by clinically achievable levels of the drug.

There is a good example of this with *Streptococcus pneumoniae*. This bacterium can become resistant to penicillin by alterations in the target site proteins, known as the penicillin-binding proteins. Depending on the alteration, these generally can confer either a moderate or major increase in resistance to penicillin. A fully susceptible bacterium may have an MIC of 0.03mg/L, whereas a moderate increase would raise this to 1mg/L and a major increase to 16mg/L. The concentrations of penicillin reaching the lung should be in excess of 1mg/L, so bacteria with this level of 'intermediate' resistance may still be susceptible to treatment; indeed, this has been recommended in some cases. On the other hand, the high-level resistance should be able to overcome the clinical levels of antibiotic.

Susceptibility levels of an antibiotic determined in the laboratory (in vitro) has rarely been correlated with the clinical efficacy (in vivo). There is a variety of reasons why a given antibiotic may behave differently in different patients: the heavier the patient, the less antibiotic may reach the site of infection as the antibiotic spreads through the body. No account is normally taken of patient size at prescribing. Thus the antibiotic may cure the same infection in a lighter patient but not in one who is obese. Also, infections behave slightly differently and each patient's immune status differs. Thus we can only make broad assumptions that as a bacterium becomes more resistant as measured in vitro, it is less likely to be successfully treated in vivo. Microbiologists continuously monitor the MICs of an antibiotic for the target bacterial populations. They usually determine the concentration of antibiotic that would inhibit 50 per cent of the population (known as the MIC_{50}). This is the median of the sensitivity and when this figure starts to rise, it means that half the bacteria isolated are showing reductions in susceptibility. A more sensitive parameter is to determine the concentration that inhibits 90 per cent of the bacteria (known as the MIC_{90}). This starts to rise as 10 per cent of the bacteria show reduced susceptibility.

Decisions are made about the continued use of antibiotics based on information such as this. Once 10 per cent of the bacteria are showing the emergence of resistance, when previously all had been sensitive, concern might be raised. When the MRSA outbreak was at its worst in British hospitals in 1998, 40 per cent of all *Staphylococcus aureus* isolations were MRSA; clearly this level of resistant bacteria is far too high. It is important to note how these figures are reached; the procedure of monitoring the levels of antibiotic resistance is called surveillance. The difficulty is that the proportion of resistance will depend on the results of which bacteria are entered into the data; for example, a hospital laboratory may measure the incidences of resistance in a particular bacterial species from lung infections. These may be completely different from the resistance levels in the same species at another site of infection. In addition, the hospital laboratory may refer a proportion of its difficult bacteria to a reference laboratory; the resistance data from the reference laboratory will show a higher proportion of resistance as the population examined has already been selected because of its resistance.

The greatest discrepancies are likely to come in general practice. The majority of infections treated with antibiotics in general practice are done so empirically, that is to say without reference to sensitivity tests. If this treatment is successful the patient will not return and, in most cases, it might be assumed that the causative bacteria were susceptible. Only if the infection does not respond to treatment might a specimen be taken and then sent to a laboratory. Suppose these cases account for 10 per cent of all infections and 50 per cent of these were found to be resistant, the surveillance data from the laboratory would show that 50 per cent of the causative bacteria were antibiotic resistant whereas the true figure might be less than 10 per cent. Surveillance is not epidemiology; only if *every* patient with a particular infection provided a specimen could we accurately assess the true level of resistance.

Despite these minutiae, there is no doubt that clinical bacteria are becoming more resistant and therapy is threatened. During the first forty years of antibiotic use, resistance was a problem but not sufficient to force the removal of individual antibiotics from clinical use. Multi-resistance emerged as a major problem in the 1980s with MRSA and was quickly followed by other bacteria that became resistant to most antibiotics. At that time, a greater demand was being made on antibiotics; previously they had primarily been used to treat acute infections that were finally resolved by the patient's immune system. The era of transplantation and aggressive cancer treatments causing neutropenia has meant that antibiotics have been used in patients with suppressed immune systems. The antibiotic alone has to keep the bacterial numbers low. This has allowed the emergence of bacteria, such as *Acinetobacter baumannii*, that have a predisposition towards resistance but can only cause infection in immunocompromised patients. The reliance on antibiotics to keep undefended patients alive has placed an intolerable burden on the antibiotics that we possess. Furthermore, during the 1980s antibiotic resistance seemed to be a problem with Gram-positive bacteria, such as MRSA or VRE. This was largely because most of the antibiotics developed up to that time, for hospital use at least, were targeted at Gram-negative bacteria, which comprised the major hospital pathogens. So the failure to tackle Gram-positive bacteria allowed resistant strains to proliferate. More recently, most new antibiotics target Gram-positive bacteria and now Gram-negative bacteria are emerging as the major resistance problem.

The challenge that this causes is much greater than it was in the 1980s, as with antiseptic treatment Gram-negative bacteria are more difficult to treat with antibiotics due to their thicker cell walls. It is proving extremely difficult to find new drugs that will be sufficiently selective to control these bacteria yet be safe enough for systemic use in humans. There are significant resistance problems to the last successful anti-Gram-negative antibiotics, ciprofloxacin, the cephalosporins, and the carbapenems, especially in hospitals. Even common hospital pathogens such as *Klebsiella pneumoniae* are becoming resistant to them all. As with MRSA, we have had to fall back on an old toxic antibiotic, colistin. Most multi-resistant Gram-negative bacteria are sensitive to colistin, but slowly resistance is beginning to develop. If it becomes firmly established, we will have almost no viable antibiotics to treat multi-resistant Gram-negative bacteria. This is likely to be a situation that is only acute in hospitals and among the immunocompromised patients; however, all hospital procedures need a risk assessment and if the risk of an untreatable infection increases for certain procedures, especially through elective surgery, they will become less feasible.

Clonality of resistant bacteria

Almost all the major multi-resistant species of bacteria are not simply the sensitive bacteria that have become resistant; rather our use of antibiotics has selected out strains that have a predisposition towards resistance. MRSA is a good example of this. There is more than 20 per cent variation within the genomes of sensitive *Staphylococcus aureus*; the species has a core genome of around 80 per cent with differences primarily found in the variable region. When the genomes of MRSA were sequenced, it was found that this diversity was lost. So what we have been witnessing with these multi-resistant bacteria is the influence of external pressures; which, of course, have been the driving force for the evolution of all bacteria. In the case of multi-resistant bacteria in general, and MRSA in particular, the selective capability of antibiotics has provided an opportunity to observe the evolution of these bacteria.

Comparison of the genomes of all sequenced strains of MRSA has been analysed with the eBURST software (Figure 21). With this it is possible to recognize relationships and identify the distances between sequences of individual strains. The MRSA strains are closely related, suggesting that their common ancestry was in the less distant past. It is possible to quantify when and how often certain genetic events took place. For example, the distinguishing feature of MRSA is the carriage of the *mec* gene that confers methicillin resistance; it appears to have migrated to MRSA strains on no more than ten occasions. However, once this gene had migrated, it was not a guarantee of success as most strains that had captured this gene have not flourished. The *mec* gene was first found in Sequence type (ST) 250 (Figure 21). It was later found in the related complexes of ST8, ST239, and ST247. The complex groups ST39, ST30, and ST36 all contain MRSA; ST36 has been the most successful in the United Kingdom and is clinically known as epidemic EMRSA-16 (Figure 21). Other features, many yet unrecognized, have been responsible for the domination of the major MRSA clones we identify today.

Similar observations have been made with the evolution of other multi-resistant bacteria, such as *Streptococcus pneumoniae* and

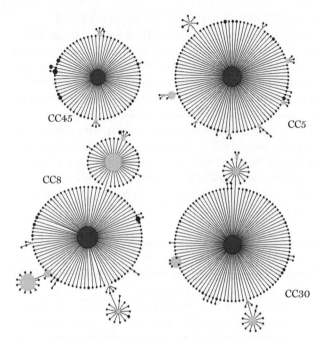

21. eBURST diagram of all the *Staphylococcus auerus* genomic sequences

Acinetobacter baumannii. In *Streptococcus pneumoniae*, the multi-resistant strains fall within specific serogroups (originally identified by their serology but now by the genomic sequences). Specific serotypes have not only been successful because of their predisposition to resist antibiotics but also by their pathogenicity. Thirteen of the most resistant serotypes have been used to make a vaccine, Prevenar 13. The widespread use of this vaccine is expected to control these serotypes; however, there is a suggestion that new, multi-resistant serotypes are emerging to fill the void created by this vaccine, so we may see the impact of an intervention, other than an antibiotic, on the evolution of these bacteria. Similarly, although there are nearly a hundred different types of *Acinetobacter baumannii*, most of the strains isolated are multi-resistant versions of just four clones, known as the worldwide clones. These clones appear much more successful than the rest. The reason for this is not clear and they possess genes, currently unknown, that promote their success. Further genomic sequencing may provide this answer in the future.

Resistance in the developing world

The introduction of new resistance mechanisms into the clinical population is a relatively rare event and most of the resistance that we witness is caused by the spread of resistant bacteria. The source of new antibiotic resistance genes is a matter of hot debate. As world travel becomes quicker and easier, it is impossible to separate individual countries and it is likely that, like certain pandemic infections, hot spots occur in some countries. It is known that the incidences of resistance in bacteria from some developing countries is much higher than in the equivalent bacteria in the industrialized world. There are reasons for this: the sanitation is often poor, so there is continual infection and reinfection. Faced with this problem, many antibiotics are obtained 'over the counter' without reference to a health-care worker, a practice only recently outlawed in some countries of the European Union. These may either be obtained in an insubstantial number of doses or the drugs themselves may be of poor quality, only containing a small proportion of the stated doses of antibiotic. In either case, the bacteria causing the infection will often be treated with sub-inhibitory concentrations, creating a powerful selection environment for resistance development. The problem will be exacerbated as the patients are likely to be undernourished, which will compromise the immune system and make it even more difficult for the antibiotic to control the infection.

There is a further issue: most of the population in developing countries lives on the land and is in close proximity to soil bacteria, a potent source of resistance genes. The mixing of these bacteria with those causing infection would allow the opportunity for the transfer of resistance genes from these soil bacteria to clinical bacteria.

Resistance in animal bacteria

There is a possibility that animals in the developing world could be a source of resistant bacteria or resistance genes; though this is not as likely as it would be with animals in the industrialized world. Most ruminants use bacteria in their gastrointestinal tract to break down the cellulose of grass and release the nutrients within. As feed for animals was modified and became richer in nutrients, the need for these bacteria decreased. They were, however, still prospering on the new feed, reducing the amount available for the host animal. In order to combat this, animals were given sub-inhibitory doses of antibiotics to suppress the growth of these commensal bacteria. These antibiotics are referred to as growth promoters and their use does significantly increase weight gain; however, the use of sub-inhibitory concentrations also promotes the development of resistance. In 1969, the Swann Report outlawed the administration of antibiotics, used in human medicine, as growth promoters. In 2006, the European Union outlawed the use of all growth promoters for fear that their use was leading to antibiotic resistance. In the United States, no ban has been placed on the use of growth promoters and they are still extensively used.

Since the Second World War, many countries, including the United Kingdom, decided to massively increase their food production. This required the intensive farming of animals, especially poultry, pigs, and fish. The massive collectivization of animals leads to rapid transmission of disease from one animal to another. In order to combat this, half the antibiotics used in the United Kingdom are used to treat infections in farm animals. They are, however, used differently from clinical practice. The treatment of human patients, for the most part, is confined to the

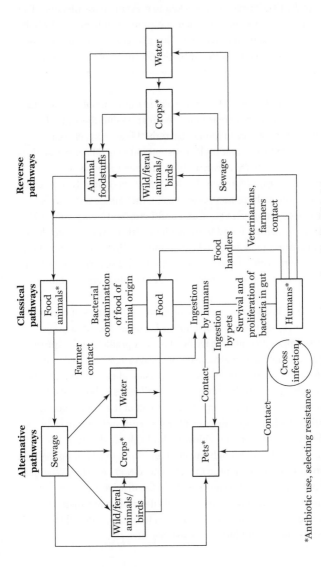

Reverse pathways

Water

Crops*

Sewage

Animal foodstuffs

Wild/feral animals/birds

Classical pathways

Food animals*

Bacterial contamination of food of animal origin

Food

Food handlers

Veterinarians, farmers contact

Farmer contact

Ingestion by humans

Ingestion by pets

Survival and proliferation of bacteria in gut

Humans*

Alternative pathways

Sewage

Water

Crops*

Wild/feral animals/birds

Pets*

Contact

Contact

Cross infection

*Antibiotic use, selecting resistance

22. Diagram showing the interactions between resistance in animals and in Man

individual and measures are taken to prevent cross-infection. This is impossible in a herd, so when one animal is infected, the whole herd is treated prophylactically. This may ensure that the animals of a herd are treated with antibiotics for much of their rearing, a practice not too dissimilar from the use of growth promoters.

It is known that many bacteria isolated from animals are resistant to antibiotics and there is much speculation that these are the source of resistance in human bacteria (Figure 22). There is certainly evidence that multi-resistant individual bacteria, such as *Salmonella enterica* Serotype Typhimurium (previously *Salmonella typhimurium*) DT 104 transferred directly from cattle to Man through contaminated meat. Other *Salmonella* strains have been reported in eggs, and multi-resistant *Campylobacter* strains have been found to transmit directly to Man. These almost always cause infections in the community. The ability of resistance genes to transmit directly to bacteria that cause serious infections in hospital is less clear and the evidence is often circumstantial.

The truth is that we usually do not know the sources of bacterial resistance and there is much unsubstantiated speculation. Furthermore, scientists have been poor at predicting the spread and even the likely mechanisms of resistance, which makes it difficult to control when a new drug is launched. Resistance now severely threatens the future use of antibiotics in hospitals. It has been suggested that during the latter half of the 20th century, antibiotics have been responsible for a ten-year increase in lifespan because of their ability to diminish the threat of premature death through bacterial infection; this is compared to a two-year increase in lifespan if all cancers were curable. Whether or not this is true, resistance without the discovery and introduction of new antibiotics will effectively make many infections more virulent, especially in hospitals, and will inevitably reverse the current increase in average lifespan.

Chapter 8
The future

The past history of bacteria has been difficult enough to uncover, let alone making valid predictions for the future.

The benefits of research into the molecular biology of bacteria

The genomic sequencing of bacteria is still in its infancy when we consider the vast number of species of bacteria on the planet. It has given us knowledge of the genes that many bacteria species require not only to survive but to progress into new niches. The more genomic sequences of each species that are completed, the more comprehensive a map of their evolution can be obtained. We have, however, less knowledge on how these genes are controlled and what stimulates them to switch on. This is crucial information because it may allow the control of pathogenic bacteria infections with drugs other than the conventional antibiotics that are used today. For instance, if the stimulus to the gene that makes a bacterium pathogenic can be inhibited, then it may be restrained without the need to kill it. This will, however, require very detailed molecular analysis.

A constraint to the successful management of patients has been the speed at which diagnostic laboratories can identify a pathogen and its antibiotic susceptibility. Currently it takes one or two days, during which time the patient has already been given therapy, which may be altered subsequently according to the test results. Molecular biological techniques are much faster and can, in theory, provide a result within an hour or two, which is within the prescribing time frame. Whereas it is now straightforward and fairly rapid to identify the pathogen by these techniques, the current constraint is the ability to translate current molecular information into predicting individual bacterial susceptibility and likely clinical success. Future extensive genomic analysis should ultimately deliver this, though when is less certain.

Bacteria as a cause of disease

In the past thirty years, there have been a number of spectacular and important discoveries where bacteria have been found to be the cause of disease. *Helicobacter pylori* and *Legionella pneumophila*, both discussed above, are two examples of where a bacterial vector of disease was unknown and unpredicted. This begs the question as to how many more bacteria, of which we currently have no knowledge, can cause disease. We have to some extent been limited by the technology we have available and this is increasingly based on molecular techniques. These are usually based on employing DNA, which will recognize known sequences; however, if the sequences have never been described before, no DNA-based technique will identify them. Therefore, it may become increasingly difficult to find new pathogens.

There are suggestions, currently without extensive proof, that bacteria play an important role in heart disease. This has been coupled with concerns that poor oral hygiene could lead to cardiovascular disease as the mouth bacteria spread into the rest of the body. These are still areas of speculation but it has been reported that *Chlamydia pneumoniae* is probably associated with heart disease as it has a coat protein that mimics a protein found in the heart muscle of mammals. The epidemiological evidence is compelling as both *Chlamydia pneumoniae* and heart disease are common in humans, but is it really cause and effect? There are clearly going to be many more questions such as this in the future.

Apparently, the greatest fear is whether a new pathogen will emerge that will have as devastating an effect as the Black Death. We have recently had scares with AIDS, SARS, and avian flu, albeit these are mainly caused by viruses. The circumstances are not the same as they were in the 6th, 14th, or 17th centuries. Our environment is cleaner, our dwellings are more self-contained, and our lives are spent in less close proximity with our neighbours than they were even before the Second World War. The spread of Spanish flu after the First World War, however, suggests that if there is widespread movement of people, an epidemic could occur. A pandemic bacterium would almost certainly have to be respiratory as other methods of transmission are largely contained with modern food production, effective sewage disposal, etc. This would mean that it would manifest itself within the cities first and then perhaps along the transport routes by rail and by air. The SARS outbreak showed how the governments of the world can react quickly once an outbreak is detected, so the chances of a major bacterial pandemic decimating the population in a vigilant society are small.

The emergence and widespread carriage of *E. coli* O157:H7 in cattle or *Aeromonas salmonicida* causing furunculosis in farmed salmon shows that when we make some change, often quite small, to an environment, it can cause some major changes in the bacteria that emerge. *E. coli* O157:H7 was unknown before farming practices changed after the Second World War. Its emergence probably results from changes in foodstuffs fed to these animals. Furunculosis does occur in wild salmon but it is rare; its emergence in farmed salmon is due entirely to the concentration of fish. Indeed, many of the new diseases in food-producing animals are related to the concentration of animals promoting spread.

If the risks from bacteria in the future are not from new, hitherto unknown, pathogens but rather confined to other animals and perhaps the environment, where are the threats? As stated earlier, we are rapidly running out of antibiotics to deal with the infections that we currently face. This threat is particularly acute with infections caused by Gram-negative bacteria. There is little likelihood of new antibiotics filling this niche in the near future, despite innumerable initiatives by governments and other grant-awarding bodies to search for new drugs. In fact, there are actually many compounds that kill bacteria but most of them are not selective and are too toxic for human systemic use. The task is not to find a new drug but rather to find one that is safe—and that is proving almost impossible.

Bacteria within the modern world

At some point in time, hopefully not until the distant future, this planet is likely to become uninhabitable for human life and perhaps also for all vertebrates. The organisms most likely to survive are bacteria. They are the most flexible organisms and evolve at a rate that will allow rapid adaptation to even more rapid changes in the environment. In the shorter term, bacteria have to survive within the modern world. The speed at which new bacteria are emerging suggests that we are still ignorant of most of the bacterial species on the planet. In the past thirty years huge numbers of new bacterial species have been identified. This has been partly because we are better able to identify and distinguish unique bacterial species, largely because we are able to analyse their DNA; however, it has also resulted from our ability to explore new environments that were previously inaccessible. A notable example is *Halomonas titanicae*, a bacterium found in the rusticles of the wreck of RMS *Titanic*, 3,800 metres below the surface of the Atlantic ocean. However, these bacteria do not operate alone and the formation of rusticles requires a consortium of other, often unknown, bacteria. If these bacteria thrive on iron, how do they normally survive at those depths in the absence of a wreck? We surely do not know.

Probably the most unlikely place to find bacteria are in volcano vents and acid springs, mainly Archaea and *Beggiatoa*, which can thrive in mud volcanoes. These environments, highly toxic to almost all animals and plants, have provided a niche for bacteria. These bacteria do not have the traditional requirements for carbon or even sunlight; they can survive on sulphur and hydrogen. One organism, *Sulfolobus solfataricus*, can thrive in temperatures as high as 88°C and in very acidic conditions. They can also survive in the volcanic vents deep under the sea.

Bacteria have been found in both the Arctic and Antarctic. About 7 per cent of the Earth's surface is covered in sea ice, and bacteria are one of the many micororganisms able to live and reproduce in it. There are four main phylogenetic groups of bacteria: the proteobacteria, the Cytophaga-Flavobacterium-Bacteroides (CFB) group, and what are known as the high and low mol per cent Gram-positive bacteria (high and low referring to the proportion of guanine and cytosine residues in their DNA); however, many novel groups are continuously being discovered, including *Polaromonas* and *Polaribacter*. Interestingly there are closely related psychrophilic or cryophilic (cold-loving) bacteria in Arctic and Antarctic sea ice, which raises the question as to how they were able to pass and survive through the tropics. The most likely explanation is that they passed between the two during one of the ice ages. The Antarctic is considerably colder as the ice rests above a huge landmass. Bacteria have been found in ice taken from the South Pole itself, a remarkable achievement of survival as they have to contend with temperatures as low as –89°C and an altitude on the ice shelf of 2,800 metres.

Many of the polar bacteria may be found as spores that could survive for up to a million years, longer than most ice ages, thus allowing bacteria from one interglacial period to thrive in the next. The bacteria would not necessarily have to be spores that survive; a study of ice from Ellesmere Island in the high Canadian Arctic revealed large numbers of bacteria within the glaciers, some of them at least 2,000 years old. Many of these bacteria, however, have been in 'suspended animation', awaiting a change in climate. Recent innovations allow deep penetration into the Antarctic ice. A pool containing unfrozen water at −10°C with a high saline concentration was found beneath 500 metres of ice in the Taylor glacier. The bacterium *Thiomicrospira arctica* was one of seventeen new types living in the pool. Although similar to some marine bacteria, these bacteria have managed to survive in the absence of either oxygen or sunlight that might have allowed photosynthesis. They managed to respire with the iron that is in the rock under the pool and appear to have survived on other living organisms cohabiting in the pool. Bacteria such as these can give an understanding of how bacteria survived the ice ages and suggest that they could survive on other planets and moons in the solar system.

Will we find new bacterial species on other planets? That is not really possible to answer at the moment. The seeking of ancient waterways and the gathering of soil samples on the surface of the planet Mars led to speculation that the planet may have had sufficient water (the prerequisite for life as we understand it) at some point. Furthermore, the discovery that there may have been bacteria-like structures, which have been given the name *Gillevinia strata*, further supported this view. However, this evidence remains inconclusive. The ALH84001 meteorite was expelled from Mars seventeen million years ago and fell onto Antarctica 11,000 years ago. Electron microscopy of the meteorite reveals some bacteria-like structures, but it is unclear whether these come from the meteor or from extended exposure to the Antarctic environment. The Shergotty meteorite, also originating from Mars, was collected in India almost immediately after it fell and examination showed that there is evidence of microbial biofilms. There is currently no firm evidence identifying bacteria on the surface of Mars itself but it is known that bacteria from the Earth have survived for some years on the surface of the Moon but no indigenous bacteria from the Moon have been identified. So it is possible for bacteria to survive outside the confines of Earth, and it is extremely likely that they do exist somewhere.

With this wealth of bacterial species, it is likely that some of them will come into direct competition with Man. The iron-devouring bacteria certainly can cause significant damage to structures such as oil rigs and bridges but their effects have been relatively slow and they are unlikely to cause us major concern. Bacteria able to consume and detoxify oil, particularly after spillages, could be particularly welcome. After the Deepwater Horizon oil spill in April 2010, an estimated 800 million litres of oil was released into the Gulf of Mexico. Approximately 25 per cent was burned or skimmed off the sea surface, leaving a vast quantity of hydrocarbons. These were quickly digested by marine bacteria such as *Alcanivorax borkumensis*. This resulted in a bacterial bloom. There was wide use of chemical dispersants, often mistakenly believed to break the oil up so it drops to the ocean floor; indeed, their main role is to break up the oil so that the bacteria can use the hydrocarbons. It was estimated that these bacteria removed over 50 per cent of the available oil. Bacteria need more than the carbon source of the oil; they require nitrogen and phosphorous and there simply was insufficient for the bacteria to remove the rest of the oil. These nutrients are usually supplied in the ocean by fluvial deposits. As time progresses, sufficient quantities of these nutrients will enter the ocean to remove all but the largest chain hydrocarbons.

There is increased interest in using bacteria such as these for boosting the yield of hydrocarbons from traditional oil wells. It is usually possible to extract 20–50 per cent of the hydrocarbons from a conventional oil well. The pressure drops and it becomes more difficult to extract the oil. Techniques such as pumping in steam or carbon dioxide under pressure have been used to increase the amount extracted. These are expensive and inefficient. In some cases, the oil deposits have been degraded by microbial action, particularly on a water–oil interface. The product has been methane and carbon dioxide. Methane is the primary constituent of natural gas and it has been proposed that bacteria should be used to convert these hydrocarbons into this convenient, easily extracted energy source. The Canadian government is funding a large project to identify and map the genes of the bacteria capable of turning oil into methane. This technology is now being targeted on previously exhausted coal mines.

Genetically modified bacteria

Forty years ago, it became possible to excise genes out of bacterial cells and splice them into other bacteria. With the appropriate genetic control machinery in place, it was then possible for these 'spliced' genes to be expressed and to confer the characteristic they carried on their new host bacterium. In essence many of these early experiments were similar to the genetic manipulations that were occurring naturally as bacteria transferred DNA in the form of plasmids from one cell to another. The technique became more sophisticated as the genes were specifically excised by restriction enzymes and spliced not into known clinical plasmids but rather into small, artificially created plasmids. These tended to be small (up to 10,000 nucleotides long) and were thus too small either to be self-transferable or to be controllable by the bacterial cell itself. These small genetically constructed plasmids were transformed into their new bacteria hosts; in other words, the DNA itself was inserted into the cell either by making the cell competent (amenable to the uptake of naked DNA) or by electroporation where the DNA inserts through holes generated by a short electric shock. The consequence of the lack of control was that, once inside, the plasmid DNA was able to replicate until there might be more than 200 copies per cell. This meant that there was much more DNA that could be transcribed and translated into protein, so the yield of the protein could be boosted 20-fold or more.

There was no theoretical limit to the genes that could be inserted into this type of plasmid; they did not necessarily have to come from other bacteria but could come from viruses, plants, animals, or even ourselves. This raised huge concerns with accusations that scientists were playing God and that the consequences of a rogue gene in the wrong bacterium could have devastating biological consequences for ourselves and the planet as a whole. A moratorium was called and rigid restrictions imposed on the type of experiments that can be performed. Nowadays, these types of genetic manipulations are under strict control; a risk assessment has to be made outlining what the potential threats are if these bacteria were to escape into the community. Unfortunately, individual countries interpret this differently and some have allowed the creation of some bacteria with potentially hazardous genes, with few or any containment restrictions.

On the positive side, there are many benefits to genetically engineered genes. An often cited example is the movement away from the use of animal-derived insulin for injection by diabetic patients. This had been the method of insulin production since the early 1920s but it was expensive and required significant purification. The further drawback was that animal insulins are not exactly the same as human ones. In the 1980s, the gene encoding human insulin was cloned into a plasmid, which was inserted into an *E. coli* strain. Culturing the bacterium in large quantities enabled high yields of insulin, which was then purified for medical use. Genetically engineered insulin now accounts for the vast majority of insulin currently used but there are concerns that the purification of the protein from the *E. coli* can allow some transfer of bacterial debris and may cause problems with allergy. Perhaps a greater concern is whether genetically engineered insulin, which is generated from the same DNA, is actually the same as natural human insulin. The three-dimensional structure of a protein is dependent on the structure of its amino acids (and thus of the DNA itself) *and* its environment, dependent on solute concentrations, temperature, and pH. The two chains of insulin are generated in bacteria and they have come from a completely different environment from that produced in animals. They are mixed to form the final molecule but it is argued that this is not identical in shape to insulin produced in animals and cannot perform the same tasks as efficiently.

This technology has been taken further where the genes encoding the human growth hormone have been inserted into bacteria and harvested directly. The previous alternative for a single dose was to extract this hormone from the pituitary glands of fifty deceased individuals. The production of Tissue Plasminogen Activator, which dissolves blood clots, has now been taken over by bacterial genetic engineering. An added advantage is that as the proteins are not produced in humans, they will not be contaminated with human viruses. This became critical when blood products were found to be contaminated with Human Immunodeficiency Virus, HIV, and many haemophiliacs became infected using contaminated factor VIII. Factor VIII is also now made by recombinant technology but in mice rather than bacteria.

The traditional use of vaccines, particularly against viral infections, is to administer either an attenuated strain that does not cause severe infection or a strain that is effectively dead and incapable of causing infection. Both rely on the fact that part of the whole virus particle is recognized as 'foreign', which becomes a target for the immune system and allows the memory T and B cells to generate specific antibodies. Both types of vaccine have disadvantages; the first is that a live virus is being administered, albeit presumed benign; the second is that a dead virus is being administered, usually at a much higher concentration, which can cause its own problems. It is argued that a much safer manufacture of vaccines would be to identify the epitopes of the virus that are the targets for the immune system and insert the DNA encoding their genes into the plasmids of bacteria. These are then cultured and the relevant epitopes (usually proteins) are harvested and comprise the vaccine. This was initially developed for Hepatitis B vaccine where the gene for the virus surface antigen, HBsAg, is cloned into a plasmid which is inserted into

E. coli. The use of the harvested HBsAg itself, rather than the whole virus, provides a safe and effective vaccine.

An alternative, if seeking to produce a vaccine for a bacterial infection, is to genetically engineer the components of the bacterium. This has proved particularly successful with *Streptococcus pneumoniae*, which we have already seen has a protective capsule; however, there are many variants of this pathogen that can cause a variety of diseases, so a genetically engineered preparation has been made to cover the main variants. In one preparation, Prevenar 13, the capsule sugars from thirteen variants (1, 3, 4, 5, 6A, 6B, 7F, 9V, 14, 18C, 19A, 19F, 23F) are cultured separately, extracted, and conjugated onto a non-toxic protein carrier, CRM197 from *Corynbacterium diptheriae*. This type of vaccination has been successful, which is encouraging as conventional vaccines against bacteria have often been less successful than those against viruses.

The use of recombinant DNA is not restricted to medical preparations. There are bacteria, for example *Bacillus thuringiensis*, that penetrate corn roots and they have been used to insert a gene that produces an insect-killing toxin, thus making corn plants resistant to detrimental insects. The genetic manipulation of the genes within *Pseudomonas syringae* has lowered the temperature at which water freezes around them. As these bacteria adhere to plants, the presence of these modified bacteria can prevent frost damage around the roots of some plants.

Synthetic bacteria

The logical progression from genetically modified bacteria would be the creation of a completely synthetic bacterium. It is possible to create genes in a nucleotide synthesizer and splice them together to form a genome. If this is inserted into a bacterial cell, which has had its DNA removed, the bacterium takes the characteristics determined by the DNA introduced into it. This act of science fiction became science fact in May 2010, when it was announced that a chemically created genome had been inserted into *Mycoplasma mycoides* by the J. Craig Venter Institute (Figure 23). This was no mean feat because large sections of DNA had to be inserted into the cell. Once inside the new bacterium, named JCVI-syn1, the DNA was able to replicate and the cells were able to divide. In other words, bacterial life had been created and satisfied a major criterion of a living organism, its ability to reproduce. The possibilities that this technology brings are huge and we are currently poised to enter a new era of synthetic and semi-synthetic bacteria that should be able to perform large numbers of different, hitherto unknown, engineering and medical tasks—the extent of which is probably only limited by our imagination.

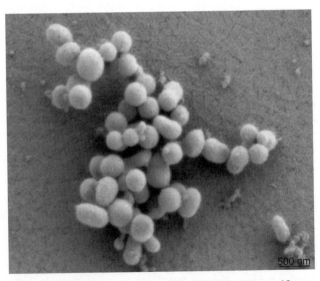

23. Micrograph of the synthetic bacterium *Mycoplasma mycoides*

There are, however, ethical and environmental arguments against this. The obvious one is the malicious creation of pathogenic bacteria, for terrorist purposes, that are more virulent and persistent than the bacteria that we currently know. It would be theoretically possible to create bacteria that had a very low dose of infectivity, so that only one or two cells could lead to a fatal infection. The bacteria could be made more resilient so that they could be combined into an explosive. Gruesome as the prospect of these bacteria might be, perhaps the greatest fear is that a bacterium would be created that would possess properties that had not been predicted or that a virulent bacterium escaped. These bacteria would not necessarily have to be directed at humans but some rogue bacterium could contaminate the environment, the domestic food-animal population, or even our fuel sources. Although our capacity to manipulate bacterial DNA has reached an extraordinary degree of sophistication, our knowledge of epidemiology and the ability to predict how new strains of bacteria will survive either in the environment or as pathogens is far less impressive. We simply do not know why some bacteria proliferate and others do not. Until we have acquired the ability to model how individual bacteria strains survive and prosper, the release of synthetically created or new and largely unknown natural bacteria has the potential, at least, to provide us with some nasty surprises unless we are very careful.

Further reading

Preface

W. B. Whitman, D. C. Coleman, and W. J. Wiebe. 'Prokaryotes; the unseen majority'. Proceedings of the National Academy of Sciences USA 95 (1998): 6578–83.

Chapter 1: Origins

E. A. Doherty and J. A. Doudna. 'Ribozyme Structures and Mechanisms'. Annual Review of Biochemistry 69 (2000): 597–615.

A. M. Poole, D. C. Jeffares, and D. Penny. 'The path from the RNA world'. Journal of Molecular Evolution 46 (1998): 1–17.

D. Jeffares and A. Poole. 'Were bacteria the first forms of life on Earth?' An ActionBioscience.org original article: http://www. actionbioscience.org/newfrontiers/jeffares_poole.html; last accessed 23 November 2012.

Chapter 2: Evolution

A. Oren. 'Biogeochemical cycles'. eLS (2008). DOI: 10.1002/9780470015902.a0000343.pub2; last accessed 23 November 2012.

R. Conrad. 'Soil microorganisms as controllers of atmospheric trace gases (H_2, CO, CH_4, OCS, N_2O, and NO)'. Microbiological Reviews 60 (1996): 609–40.

R. Calendar. *The Bacteriophages*. (New York: Oxford University Press, 2006).

Chapter 3: Discovery

P. de Kruif. *Microbe Hunters*. (Orlando, FL: Mariner Books, 2002).

S. G. B. Amyes. *Magic Bullets: Lost Horizons*. (London: Taylor and Francis, 2001).

D. H. Crawford. *Deadly Companions*. (Oxford: Oxford University Press, 2007).

J.-P. Butzler. '*Campylobacter*, from obscurity to celebrity'. Clinical Microbiology and Infection 10 (2004): 868–76.

Chapter 4: Environment and civilization

L. Gram, L. Ravin, M. Rasch, J. B. Bruhn, A. B. Christensen, and M. Givskov. 'Food spoilage—interactions between food spoilage bacteria'. International Journal of Food Microbiology 78 (2002): 79–97.

P. Kindstedt. *Cheese and Culture: A History of Cheese and its Place in Western Civilization*. (White River Junction, VT: Chelsea Green, 2012).

C. A. Wilson. *Waste Not, Want Not: Food Preservation in Britain from Early Times to the Present Day*. (Edinburgh: Edinburgh University Press, 1991).

Chapter 5: Bacterial pathogenesis

C. A. Mims, A. Nash, and J. Stephen. *Mims' Pathogenesis of Infectious Disease*. (London: Elsevier, 2001).

B. Henderson, S. Poole, and M. Wilson. 'Bacterial modulins: a novel class of virulence factors which cause host tissue pathology by inducing cytokine synthesis'. Microbiology Reviews 60 (1996): 316–41.

H. Schmidt and M. Hensel. 'Pathogenicity islands in bacterial pathogenesis'. Clinical Microbiological Reviews 17 (2004): 14–56.

Chapter 6: Antibiotics

S. G. B. Amyes. *Antibacterial Chemotherapy*. (Oxford: Oxford University Press, 2010).

F. Ryan. *Tuberculosis: The Greatest Story Never Told*. (Bromsgrove: Swift Publishers, 1992).

C. Walsh. *Antibiotics: Actions, Origins, Resistance*. (Washington, DC: American Society for Microbiology Press, 2003).

E. M. Scholar and W. B. Pratt. *The Antimicrobial Drugs*. (New York: Oxford University Press, 2000).

M. Wilson and D. Devine. *Medical Implications of Biofilms*. (Cambridge: Cambridge University Press, 2003).

Chapter 7: Antibiotic resistance

S. B. Levy. *The Antibiotic Paradox: How the Misuse of Antibiotics Destroys their Curative Powers*. (Cambridge, MA: Perseus Publishing, 2002).

R. Finch, P. Davey, M. Wilcox, and W. Irving. *Antimicrobial Chemotherapy*. (Oxford: Oxford University Press, 2012).

K. Bush and J. F. Fisher. 'Epidemiological expansion, structural studies, and clinical challenges of new beta-lactamases from Gram-Negative bacteria'. Annual Review in Microbiology 65 (2011): 455–78.

R. Bonomo and M. Tolmasky. *Enzyme-mediated Resistance to Antibiotics*. (Washington, DC: American Society for Microbiology Press, 2009).

I. Phillips, M. Casewell, A. Cox, B. De Groot, C. Friis, R. Jones, C. Nightingale, R. Preston, and J. Waddell. 'Does the use of antibiotics in food animals pose a risk to human health? A critical review of published data'. Journal of Antimicrobial Chemotherapy 53 (2004): 28–52.

Chapter 8: The future

R. Rappuoli. 'From Pasteur to genomics: progress and challenges in infectious diseases'. Nature Medicine 10 (2004): 1177–85.

L. D. Rotz and J. M. Hughes. 'Advances in detecting and responding to threats from bioterrorism and emerging infectious disease'. Nature Medicine 10 (2004): S130–S136.

R. Maresin, F. Schinner, J.-C. Marx, and C. Gerday. *Psychrophiles: From Biodiversity to Biotechnology*. (Berlin: Springer-Verlag, 2008).

D. G. Gibson, J. I. Glass, C. Lartigue, V. N. Noskov, R.-Y. Chuang, M. A. Algire, G. A. Benders, M. G. Montague, L. Ma, M. M. Moodie, C. Merryman, S. Vashee, R. Krishnakumar, N. Assad-Garcia, C. Andrews-Pfannkoch, E. A. Denisova, L. Young, Z.-Q. Qi, T. H. Segall-Shapiro, C. H. Calvey, P. P. Parmar, C. A. Hutchison, H. O. Smith, and J. C. Venter. 'Creation of a bacterial cell controlled by a chemically synthesized genome'. Science 329 (2010): 52–6.

Index

JOIN OUR COMMUNITY

www.oup.com/vsi

- Join us online at the official Very Short Introductions **Facebook** page.
- Access the thoughts and musings of our authors with our online **blog**.
- Sign up for our monthly **e-newsletter** to receive information on all new titles publishing that month.
- Browse the full range of Very Short Introductions online.
- Read **extracts** from the Introductions for free.
- Visit our library of **Reading Guides**. These guides, written by our expert authors will help you to question again, why you think what you think.
- If you are a teacher or lecturer you can order inspection copies quickly and simply via our website.

Visit the Very Short Introductions website to access all this and more for free.

www.oup.com/vsi

ONLINE CATALOGUE

Very Short Introductions

Our online catalogue is designed to make it easy to find your ideal Very Short Introduction. View the entire collection by subject area, watch author videos, read sample chapters, and download reading guides.

http://fds.oup.com/www.oup.co.uk/general/vsi/index.html

Epidemiology
A Very Short Introduction
Rodolfo Saracci

Epidemiology has had an impact on many areas of medicine; from discovering the relationship between tobacco smoking and lung cancer, to the origin and spread of new epidemics. However, it is often poorly understood, largely due to misrepresentations in the media. In this *Very Short Introduction* Rodolfo Saracci dispels some of the myths surrounding the study of epidemiology. He provides a general explanation of the principles behind clinical trials, and explains the nature of basic statistics concerning disease. He also looks at the ethical and political issues related to obtaining and using information concerning patients, and trials involving placebos.

www.oup.com/vsi

HIV/AIDS
A Very Short Introduction
Alan Whiteside

HIV/AIDS is without doubt the worst epidemic to hit humankind since the Black Death. The first case was identified in 1981; by 2004 it was estimated that about 40 million people were living with the disease, and about 20 million had died. The news is not all bleak though. There have been unprecedented breakthroughs in understanding diseases and developing drugs. Because the disease is so closely linked to sexual activity and drug use, the need to understand and change behaviour has caused us to reassess what it means to be human and how we should operate in the globalising world. This *Very Short Introduction* provides an introduction to the disease, tackling the science, the international and local politics, the fascinating demographics, and the devastating consequences of the disease, and explores how we have — and must — respond.

'It won't make you an expert. But you'll know what you're talking about and you'll have a better idea of all the work we still have to do to wrestle this monster to the ground.'

Aids-free world website.

THE HISTORY OF MEDICINE
A Very Short Introduction
William Bynum

Against the backdrop of unprecedented concern for the future of
health care, this Very Short Introduction surveys the history of
medicine from classical times to the present. Focussing on the
key turning points in the history of Western medicine, such as the
advent of hospitals and the rise of experimental medicine, Bill
Bynum offers insights into medicine's past, while at the same time
engaging with contemporary issues, discoveries, and
controversies.